The Rise of the U.S. Environmental Health Movement

The Rise of the U.S. Environmental Health Movement

Kate Davies

ROWMAN & LITTLEFIELD PUBLISHERS, INC.
Lanham • Boulder • New York • Toronto • Plymouth, UK

Published by Rowman & Littlefield Publishers, Inc.
A wholly owned subsidiary of The Rowman & Littlefield Publishing Group, Inc.
4501 Forbes Boulevard, Suite 200, Lanham, Maryland 20706
www.rowman.com

10 Thornbury Road, Plymouth PL6 7PP, United Kingdom

British Library Cataloguing in Publication Information Available

Library of Congress Cataloging-in-Publication Data

Davies, Kate, 1956- .
Rise of the U.S. environmental health movement / Kate Davies.
p. cm.
Includes bibliographical references and index.
ISBN 978-1-4422-2137-6 (cloth : alk. paper) — ISBN 978-1-4422-2138-3 (electronic)
1. Environmental health—United States—History. 2. Pollution—Environmental aspects—United States. 3. Environmentalism—United States—History. I. Title.
RA566.3.D38 2013
613'.1—dc23
2013000229

∞™ The paper used in this publication meets the minimum requirements of American National Standard for Information Sciences Permanence of Paper for Printed Library Materials, ANSI/NISO Z39.48-1992.

Printed in the United States of America

“The earth does not belong to man: man belongs to the earth. This we know. All things are connected like the blood, which unites one family. All things are connected. Whatever befalls the earth befalls the sons of the earth. Man did not weave the web of life; he is merely a strand it. Whatever he does to the web, he does to himself.” —Attributed to Chief Seattle

Contents

Foreword

Elise Miller

Every other week my eight-year-old son goes to an after school chemistry club with his buddy—meaning they spend an hour with a retired science teacher who is passionate about passing on his knowledge to the next generation. His "lab" is something out of the early twentieth century, containing rows of dusty bottles with handwritten labels identifying specific chemicals or simply warning of danger, wires, and gadgets that cover shelves and spill out of boxes, as well as an antique wood stove that never quite gets the room warm. On their first day, he introduced the periodic table, and since then, he has shown them everything from explosive chemical reactions to electromagnetic fields. The boys love him and eagerly anticipate their biweekly visits.

Given the twenty years I have worked on environmental health issues, I observe this class both with exhilaration—watching these young kids get excited about learning about chemistry—and with discomfort—wondering if they are sufficiently protected from hazardous exposures in his lab, knowing all too well that they can undermine healthy child development.

In the larger picture, I contemplate whether our society will continue to let children grow up breathing, eating, and drinking toxins that can lead to chronic diseases and disabilities, or whether it will choose to prioritize prevention and develop the safest possible chemicals. Given that everyone is exposed to hundreds of contaminants every day, particularly those who live in poorer neighborhoods where the toxic burden is often far greater, the fundamental question is what can we do now so that our kids and grandkids don't have to worry about harmful exposures wherever they live, play, study, and work? What steps can we take to ensure they don't suffer from preventable diseases and that they reach their full potential?

Kate Davies' book *The Rise of the U.S. Environmental Health Movement* gets to the heart of these questions and why they should concern everyone who cares about their own health and the health of their families and communities. She provides a persuasive analysis of what originally compelled our culture to undertake this toxic experiment and the social movement that is emerging across various sectors to address it. Given the growing scientific literature that associates environmental contaminants with the rising rates of so many health problems—from learning disabilities to obesity to cancer—this book is essential not only to read, but to digest and share widely.

I met Kate when she first moved to the Seattle area about ten years ago. I immediately recognized her deep-rooted experience and leadership in different facets of the burgeoning environmental health movement, and asked her to serve on the board of the national Institute for Children's Environmental Health (ICEH), which I founded and directed. Her contributions in that capacity added significantly to the Institute's productivity and effectiveness. Over time, we not only became collaborating colleagues, but close friends as well.

In 2009 I merged ICEH with Commonweal, a health and environmental research institute and educational center, and became director of the Collaborative on Health and the Environment (CHE), one of its largest programs. CHE, founded in 2002, was the first major national organization to reach out to a wide range of groups representing people affected by environmentally related diseases and disabilities and engage them in discussion on the emerging science on environmental contributors to various diseases and disabilities. CHE is now an international partnership committed to strengthening the scientific and public dialogue on the environmental factors linked to chronic disease and disability and catalyzing multistakeholder collaborative initiatives to address these issues. With over forty-five hundred members in all fifty states in the United States and seventy-nine countries, CHE continues to be successful in encouraging institutions and whole sectors to put environmental health at the center of their decision making.

When Kate, who is a valued member of CHE's broad network, decided to write this book, I had the privilege of serving as her sounding board at various times. Over the several years it took her to research, write, and now publish this volume, I found myself impressed by her dedication to sifting through and reflecting on the vast amount of information relevant to this project. The result of her tireless efforts is an unprecedented, invaluable, and timely synthesis of the historical threads and current patterns that make up the fabric of the U.S. environmental health movement.

One reason *The Rise of the U.S. Environmental Health Movement* is a "must read" for anyone interested in environmental health is that it doesn't cover only one sector—such as health sciences or public health policy. Nor does it analyze the movement from simply one angle—such as the emerging science or active nonprofit organizations or environmental justice. Instead, Kate's book is particularly notable because it provides a comprehensive assessment of multiple interactions among multiple sectors from multiple perspectives. That said, Kate wisely does not try to cover every field and topic that may be relevant to environmental health, such as access to health care, and social and emotional stressors. Instead, she notes the complexity of the issues connected to this field, while staying focused primarily on toxins. That seems like the right scope for this already ambitious volume.

Another reason this book makes such an intelligent contribution to the field is that it's useful not only for those of us immersed in these issues every day, but for those who are completely unschooled in environmental health. In other words, by placing the myriad facets of this movement in context, Kate helps us all to understand the origins and evolution of environmental health problems so that we can take wiser action. Along these lines, she covers not only the "thirty-thousand-foot" view, but makes this material accessible on the personal level as well, providing references for everyday choices about consumer products and what we can do individually to make a positive difference in our families and communities.

Finally, *The Rise of the U.S. Environmental Health Movement* is a finely balanced and fair-minded account of how this movement came to be and what it will take to execute the sea change we need to fully protect public health. This kind of neutral articulation is not easy to come by these days when perspectives are often deeply polarized and divisive.

Some of those in the U.S. environmental health movement may, of course, view some aspects of Kate's analysis differently. My sense is that any points of divergence will only serve to enrich discussions among those of us who see environmental health as a major part of our life's work. Furthermore, by continuing to engage each other as well as widening the circle of those who understand how environmental health is relevant to all of us, we have an opportunity to effectively leverage the kind of fundamental shift in thinking and practice we need at all levels of society.

When I finished reading this book, I was reminded of the extraordinary array of talented colleagues committed to these issues—not only on a pragmatic, strategic level, but on the heart level. Though my son's generation will likely bear the burden of our toxic legacy, the environmental health movement can also help imbue him and his peers with a conscious understanding that humanity is not separate from all other species and the earth—nor are we simply stewards. We are, in fact, kin. And as kin we have the moral and ethical responsibility to act in ways that promote and protect all life, not simply our own kind. By acting from that knowledge, which I believe Kate's book inspires us to do, we have an opportunity and an obligation to work collectively and scale the high hurdles still ahead in order to ensure our world is the healthiest it can be for generations to come.

Elise Miller, M.Ed., is director and cofounder of the Collaborative on Health and the Environment (CHE), a project of Commonweal. Ms. Miller founded and directed the Institute for Children's Environmental Health for ten years before merging ICEH with Commonweal in 2009. From 1993 to 1998 she served as the founding executive director of the Jenifer Altman Foundation, a small private foundation with interests in environmental health and justice.

Acknowledgments

Writing a book is like building a social movement; it requires the ideas, time, and support of many people. This book is no exception. Without the help of my friends, family, and colleagues, as well as the wonderful people who have dedicated themselves to protecting environmental health in the United States, it would not have been possible.

First, I would like to acknowledge the Sustainable Path Foundation, whose financial support enabled me to write an early version of this book. I would also like to express deep thanks to my friends Elise Miller, director of the Collaborative on Health and the Environment, and Linda Park and Nan McKay. Their unwavering interest and encouragement sustained me, especially when I felt like giving up. I am very grateful to my colleagues at the Center for Creative Change at Antioch University Seattle, especially Don Comstock. Their ideas were invaluable in helping me to think about progressive social change and environmental health.

Perhaps most importantly, I would like to acknowledge everyone in the U.S. environmental health movement who contributed to this book by providing me with their insights, suggestions, and comments, especially Laura Abulafia, Mike Belliveau, Dave Bennett, Ann Blake, Aimee Boulanger, Joy Carlson, Lin Kaatz Chary, Richard Clapp, Lindsay Dahl, Gary Cohen, Carol Dansereau, Jeannie Economos, Nancy Evans, Ken Geiser, Steve Gilbert, Michael Green, Renee Hackenmiller-Paradis, Lauren Heine, Richard Jackson, Denny Larson, Rich Liroff, Stacy Malkan, Sarah Miller, Peter Montague, Anne Rabe, Carolyn Raffensperger, Judy Robinson, Cindy Sage, Jennifer Sass, Barbara Sattler, Ted Schettler, Chloe Schwabe, Alexandra Gorman Scranton, Mike Schade, Joel Tickner, David Wallinga, Michael Wilson, and the anonymous peer reviewers.

Finally, my deepest gratitude to my son Jonathan and my husband George for sharing this journey with me.

Introduction

I was thirty-four years old when my son was born. As manager of Toronto's Environmental Protection Office in Canada, I had already been writing, speaking, and teaching about environmental health for many years. From a scientific perspective, I knew all about the pollutants present in my drinking water, the lead emissions from the secondary smelter located inside the city limits, and the contaminated soil underneath much of the downtown core. But what I didn't know about were my feelings. I didn't know what it would feel like as I held my son to my breast and nursed him—nursed him my milk that was likely contaminated with PCBs, dioxins, DDT, trichloroethylene, perchlorate, and a cocktail of other toxic chemicals. As I held him close and fed him in the most intimate way possible, I had to confront the horrible reality that I was passing on my toxic legacy to him—a completely innocent and vulnerable child. At the same time as he was ingesting the essential proteins, fats, carbohydrates, antibodies, and other nutrients in my milk, he was also getting exposed to the pollutants in my body. This realization shook me deeply and made me understand environmental health in an entirely new way. It was no longer just an abstract, scientific issue; it was intensely personal and real. As a result of this powerful experience, I redoubled my professional work, holding it in a deeper, more meaningful way. Environmental health became much more than just a scientific matter; it became an ethical, moral, and spiritual issue.

So what exactly is environmental health and why is it a problem?

ENVIRONMENTAL HEALTH

Defining environmental health is actually more complicated than it might appear. To some, it's the health of the environment, including wildlife, plants, and the planet's life support systems. To others, it's all the environmental factors that affect human health and well-being. Consistent with this understanding, the World Health Organization (WHO) defines environmental health as "all the physical, chemical, and biological factors external to a person . . . that can potentially affect health."[1] These two ways of thinking about environmental health are quite different; one foregrounds the health of the environment, and the other foregrounds human health. But in reality, they're complementary because human health depends on the health of the environment. Unless the planet is healthy, people can't be truly healthy. As cultural historian Thomas Berry said: "You cannot have well humans on a sick planet."[2]

With this in mind, most people think of environmental health in the same way as the World Health Organization—the factors in the environment that affect human health. So what environmental factors affect our health? Well, just about everything, including air, water, food, soil, climate, and other living things. But those are only the factors in the natural environment. It's also important to consider how the built environment affects us. The average American spends about 90 percent of her/his time indoors,[3] so the built environment is important too. This includes our cities and communities, workplaces, homes, schools, and other public places. Put it all together, and the extremely broad scope of environmental health becomes obvious.

But why should you be concerned about environmental health? After all, health has improved more over the past century than at any other time in human history. Since 1900 infant mortality in the United States has fallen by a whopping 90 percent and life expectancy has increased by almost 40 percent. Similar patterns hold true globally, with life expectancy more than doubling and infant mortality decreasing by about half.

On the surface, these achievements appear quite extraordinary. But dig a little deeper, and it becomes obvious that they come with a huge price tag: the planet's sustainability. The health and social services which have made these health improvements possible have been paid for by economic development that has exploited the planet's natural resources. In the past fifty years, humankind's ecological footprint has more than tripled, so that every year we are consuming 50 percent more resources than the earth can regenerate.[4] We are no longer living off the planet's interest; we are depleting its capital and destroying the planet's life support systems. In effect, recent improvements in human health have been purchased at the expense of the earth's health.

If we continue business as usual, by the early 2030s humankind will need to consume more than two planets' worth of natural resources every year just to maintain our current standard of living. Obviously, this is impossible. Unless we change our ways, it will become increasingly difficult to sustain human health and well-being and declines will be inevitable. Indeed, we are already witnessing the early warning signs, and predictions for the future are grim.

Globally, supplies of freshwater, fish, wood, oil, and many other natural resources are already dwindling and will get progressively scarcer, causing increasing health problems as well as social, economic, and political upheaval. Over eighty countries already face severe water shortages, and by 2025 it is predicted that about half of the world's population will not have enough to drink.[5] Between 1995/97 and 2009 the number of hungry people in the world increased from almost eight hundred million to well over a billion,[6] and food security has become a major international issue. Aggravating the situation, about 50 percent of the world's fisheries are already fully exploited and another 25 percent have been depleted beyond their ecological safety limits,[7] leaving many people without jobs and an important source of food. Looking at nonrenewable energy, many commentators agree that we have already reached "peak oil," the maximum rate of crude oil extraction, although others point to small increases in production rates in recent years. Whatever the truth, global oil reserves are undeniably shrinking while demand continues to rises. As a result, prices will continue to rise, causing social instability and threatening the health and well-being of many countries.

Meanwhile, at the back end of the global economic production machine, we are poisoning the water we drink, the air we breathe, and the food we eat. Pollution is now ubiquitous; there is nowhere on the planet that is uncontaminated. Some of the highest levels of pollution are in the Arctic, many thousands of miles away from any direct sources. We now know that the Great Pacific Garbage Patch, first discovered in 1997, is not a unique phenomenon; there is another one located in the North Atlantic. According to the UN Environment Programme there are about thirteen thousand pieces of plastic litter floating in every square kilometer of ocean,[8] equivalent to almost fifty thousand pieces of plastic per square mile.

Perhaps most tragically, natural resource exploitation and pollution are combining to cause the largest species extinction since the dinosaurs, threatening the existence of nearly one-quarter of all mammalian species and about one in eight bird species. Scientists say that the current rate of biodiversity loss is between one thousand and ten thousand times greater than the natural rate.[9] This issue is of particular concern for human health because most medicines, including all antibiotics and most anti-cancer drugs, come from plants, animals, or microbes, and because changes in biodiversity increase the range and severity of many environmentally transmitted diseases, such as malaria and schistosomiasis.

And then there is global climate disruption. More than seventy thousand people died because of the record heat wave that scorched Europe in 2003[10] and eleven thousand died in Moscow alone during the Great Russian Heat Wave of 2010.[11] As temperatures soar, many more heat-related deaths are anticipated. But rising temperatures aren't the only concern; increasing smog levels in cities will result in even more deaths in the years to come. Already, 1.3 million people die from outdoor air pollution each year,[12] and this number will continue to rise. As well, significantly higher rates of many environmentally transmitted diseases, including cholera, typhoid, and malaria, are likely. In some places, they're already a reality. For the first time ever, malaria has spread into the highland regions of East Africa.[13] Warmer, wetter conditions are providing new breeding grounds for the mosquitoes that spread this potentially fatal disease. Summarizing the effects, in 2009 the Global Humanitarian Forum stated: "every year climate change leaves over 300,000 people dead, 325 million people seriously affected, and economic losses of US$125 billion. Four billion people are vulnerable, and 500 million people are at extreme risk."[14]

There's no doubt that the global burden of environmentally related illness is already staggering and will get even larger. The World Health Organization (WHO) has reported that nearly one-quarter of all human disease is due to poor environmental quality[15]—almost half of all asthma (44 percent), about one-fifth of all cancers (19 percent), about one-sixth of all cardiovascular disease (16 percent), and one-twentieth of all birth defects (5 percent). Tragically, the proportion is even higher for children—about one-third. To add to this, the World Health Organization has attributed 4.9 million of the deaths that occurred in 2004 to environmental exposure and selected chemicals.[16] This represents 8.3 percent of the total deaths and 5.7 percent of the total burden of disease worldwide. This is more than the combined deaths from HIV/AIDS, tuberculosis, and road traffic accidents. Not even including the number of illnesses caused by environmental exposures, this number reveals the enormity of environmental health problems.

Although you may not be concerned about what's happening at the global level, the situation in the United States is equally alarming. Chances are, you or someone you love has an environmentally related illness. Rates of chronic diseases are skyrocketing, and more than one hundred million men, women, and children—a full one-third of the population—now suffers from one or more long-lasting health problem. Many of these illnesses, including asthma, cancer, cardiovascular disease, infertility, developmental and learning disabilities, Alzheimer's, and Parkinson's, have been linked with the environment. In fact, environmental toxicants have been associated with over two hundred diseases and disabilities.[17]

The epidemic of environmentally related illness in the United States is not surprising. Almost half of the population—about 128 million people—is breathing polluted air and lives in a county that exceeds one or more of the national air quality standards,[18] one in four Americans live within four miles of a toxic waste dump,[19] and about 49 million people are drinking water with illegal concentrations of chemicals, radioactive substances, or dangerous bacteria.[20] We all eat food containing toxic chemicals—even those who eat organic.

The truth is that everyone is exposed to toxic chemicals and other environmental hazards every day of their lives. We cannot escape from the chemical environment in which we live. There are about eighty thousand chemicals in use in the United States and about one thousand new ones are added each year. Not only is the U.S. chemical industry the largest in the world, polluters discharge millions of pounds of wastes into the environment 24/7. Toxic chemicals are present in the air we breathe, the water we drink, the food we eat, and the consumer products we buy.

We know we are exposed because everyone has environmental contaminants in their bodies—you, me, our children and grandchildren, our neighbors and friends, and people who live on the other side of the country who we will never know. Even newborn babies. In fact, virtually everyone alive—all seven billion of us—have toxics in their bodies. Present in our hair, blood, fat, bones, teeth, and other tissues, they provide incontrovertible proof of human exposure. And even worse, most chemicals have not been adequately tested for their safety. Whether we like it or not, we are all part of a massive, unprecedented, and uncontrolled scientific experiment, and the early results are extremely troubling.

But it's not just the scale and severity of the harm that are so scary; it's also the financial costs. Expenditures on environmentally related diseases and disabilities are enormous and represent a significant drain on the U.S. economy. Whether it's the cost of medical and nursing care, lost productivity, premature deaths, or health and social services, the numbers quickly add up. One study calculated the costs of seven environmentally related diseases in children at $76.6 billion a year.[21] Another estimated the health costs of air pollution at $53 billion a year.[22] And a third found that the health costs of six climate disruption-related events in the United States totaled $14 billion.[23] There's no doubt that environmentally related diseases and disabilities cost U.S. taxpayers and businesses hundreds of billions of dollars a year. And that's a lot more than just small change.

Quantifying the economic costs of environmentally related diseases and disabilities is important because it gets attention. By speaking the language of dollars and cents, activists can influence policy and get a seat at the decision-making table. But it's also important to remember that the economic costs of environmentally related illness can never include their emotional toll. The pain and suffering experienced by victims, their families, and their communities are unmeasurable and as a result, these costs are rarely considered by legislators. Policy makers would do well to remember Albert Einstein's pithy comment: "Not everything that counts can be measured." The tragic irony is that most environmentally related diseases and disabilities and their costs could be prevented. By reducing pollution, the United States could reduce the rates of many. This would be good for the economy, good for the environment, and good for human health.

It's easy to blame the chemical industry for environmental health problems or to criticize the government for allowing pollution to happen, but the real culprit is our society's core beliefs about humankind's relationship with the natural world. Although we are largely unconscious of them, these beliefs are deeply embedded in our collective psyche and influence how we think and act. One of them is anthropocentrism—the idea that our species is the central and most important species. Ever since the days of ancient Greece, we have presumed that *Homo sapiens* is more advanced and ontologically separate from any other form of life. But as Barry Commoner said more than forty years ago in his *Four Laws of Ecology*, "Everything is connected to everything else. There is one ecosphere for all living organisms and what affects one, affects all."[24] The truth is that the belief that we are separate from and superior to nature is largely responsible for the ecological crisis and many environmental health problems. We need to remember that we are affected by the environment because we are part of the web of life.

THE U.S. ENVIRONMENTAL HEALTH MOVEMENT

The environmental health movement is the only social movement to fully understand the nature of the relationship between humankind and the environment. It alone is based on the awareness that human health and well-being depend on the environment. Informed by the knowledge that our species would quickly perish without access to safe drinking water, clean air, nutritious food, and health-supporting surroundings, the environmental health movement acknowledges our complete reliance on the earth.

Although its roots go back more than two thousand years, the environmental health movement was born in 1978 when a group of community residents came together to protest the health effects caused by a local toxic waste dump in Love Canal, New York. Since then, it has grown into a national and international phenomenon. Today, citizens everywhere are demanding that governments and corporations stop pollution and protect their health. From the Americas to Africa and from Europe to Asia, people are coming together. Not only to talk about how their health has been affected by the environment, but to call for change. And how large is this movement? About ten thousand environmental health organizations and people are listed on WISER Earth, a worldwide social networking website for progressive change. More than forty-five hundred members in seventy-nine countries and all fifty states form the United States-based Collaborative on Health and the Environment. And these numbers are just the tip of the iceberg.

Placing human health unambiguously at the center of concern distinguishes the environmental health movement from other social movements, including the environmental and environmental justice movements. Environmentalism situates the environment at its core, rather than humankind. Seeking to protect pandas and polar bears, wilderness and wetlands, it foregrounds the natural world rather than human beings. And although environmental justice focuses on our species, it foregrounds human rights, rather than health. To be sure, the U.S. environmental health movement grew up in the shadow of the environmental movement and alongside the environmental justice movement and is closely allied with both. It will always have many friends and colleagues in these kindred movements, but its primary focus on human health makes it unique.

Although environmental health is an extremely broad subject, the movement has targeted a single issue—opposing toxic chemicals—for most of its lifetime. Whether it's been protesting pollution or lobbying for legislation, the U.S. environmental health movement has worked tirelessly to oppose toxic chemicals. For more than three decades, activists have tried to stop the manufacture, use, and disposal of these harmful substances. And they've had considerable success. By calling attention to the fact that human health and well-being are threatened by toxic chemicals, the U.S. environmental health movement has played a leading role in strengthening government policies and programs, changing corporate and business practices, and influencing the behavior of individuals. But more importantly, it's reminding Americans that human health and well-being depend on the environment—that we aren't separate from it. In doing so, the U.S. environmental health movement is pointing to values and beliefs that may help to resolve the ecological crisis. This is because how we think about the human-environment relationship will influence our decisions and actions as a society. To put it succinctly, the U.S. environmental health movement is not only benefitting human health and well-being, it's also transforming our culture's belief in anthropocentrism, thereby changing the way society behaves.

Opposing toxic chemicals has been an excellent strategic issue because these substances are everywhere. Found in indoor and outdoor air, drinking water, surface water and groundwater, urban and agricultural soil, meat, fruits, vegetables, dairy products and other foods, and our very own bodies, they're quite literally omnipresent. In this way, work on toxic chemicals touches on many other environmental health concerns. In recent years, the U.S. environmental health movement has branched out further and begun to work on nanotechnology, electromagnetic fields, food, fossil fuels, and green building, and some of its leaders are arguing for an even more comprehensive approach that integrates all the factors that affect human health. Called "ecological health," it advocates looking at the interactions between genetics, lifestyle, race, ethnicity, socioeconomic status, and environmental factors.

Whether or not the whole movement moves in this direction is still an open question, however, there's no doubt that the concept of ecological health is broadening its scope.

The increasing breadth of issues is mirrored in the diversity of people drawn to the U.S. environmental health movement. Scientists and researchers, health professionals and patients, attorneys and activists, teachers and students, and most importantly, concerned citizens—all are attracted to its cause. They intuitively understand that health depends on the environment. They know that everyone shares its water, air, and food and that the environment is a common resource for all. Clean or contaminated, our individual and collective health depends on it. This makes environmental health a great social equalizer. Transcending economic, political, and cultural divides, this movement speaks to everyone. Because we all want to live in healthy environments, it can bring together many different voices.

But does the range of issues and supporters disqualify environmental health as a social movement? Absolutely not. Cultural historian Sidney Tarrow defines social movements as "collective challenges (to society's power-holders) by people with common purposes and solidarity in sustained interactions with elites, opponents and authorities."[25] The U.S. environmental health movement satisfies all of Tarrow's criteria. Its common purpose unifies every group and action. Its social, economic, and political critique offers a compelling collective challenge to society's power-holders. And its thirty-five-year track record attests to its staying power. These are the *bona fides* of environmental health as a social movement.

However, it's also true that the very nature of social movements is changing, and environmental health is no exception. Like other twenty-first-century movements, it lacks some of the hallmarks of earlier ones. For instance, the U.S. environmental health movement has few iconic leaders and spokespeople, and it doesn't have a hierarchy or a centralized power structure. But these features are turning out to be some of its strongest assets. The lack of iconic leaders means that anyone can step up and assume responsibility; it is not a leaderless movement, rather it is one of collective leadership and collaboration. Similarly, the lack of hierarchy and a centralized power structure means that the movement is flexible and can adapt quickly to changing circumstances. As Ken Geiser, professor at the University of Massachusetts at Lowell, told me, "One of the environmental health movement's strengths is that there is no central command."[26] Another way to think about its flexibility and fluidity is to see the environmental health movement, and other new movements, as verbs rather than nouns. If viewed as unfolding social processes, they can be seen as emergent phenomena rather than fixed entities. Like the new land constantly being created by Hawaii's volcanoes, no one can foresee its eventual topography. But unlike the situation on Hawaii, we can do more than merely observe the appearance of a new landscape. We can become active participants and help to build a healthy, just, and sustainable world.

MY BACKGROUND

I have written this book based on my experience and knowledge of the U.S. environmental health movement; I am not a dispassionate observer but rather a participant, and like many, my commitment is a consequence of tragic personal experience. In 1965, when I was eight years old, my mother was diagnosed with Hodgkin's lymphoma, a form of cancer. She was given less than a year to live. By some miracle she survived, only to be diagnosed with breast cancer some twenty years later. She survived this too, but in 1995 she developed a rare T-cell lymphoma. She died in 2007, after fighting three different types of cancer for over forty years.

My mother's illnesses influenced me profoundly. As a child, I wanted to become a doctor so I could make her better, but as the physicians failed to cure her, I became more interested in how cancer could be prevented. To find out more, I decided to study biochemistry. After completing a bachelor's degree in 1978, I went on to earn a doctorate at Oxford University. During this time, I became convinced that toxic chemicals and radiation played a role in this terrible disease—a realization that led me to join the environmental health movement.

After working for Greenpeace on the health problems caused by uranium mining and nuclear weapons testing, I was hired by the City of Toronto in Canada to work on environmental health policy. Then in 1984, I set up and became the manager of its Environmental Protection Office—the first local government environment office in Canada. In this role, I developed policies and programs on a wide range of environmental health issues, including air and water quality, pesticides, the right to know, and land use planning. I became active in provincial, federal, and international environmental health policy and was appointed to several government committees and nonprofit boards, including the Ontario Pesticides Advisory Committee, the Canadian Environmental Assessment Research Council, the Canadian Environmental Law Association, and the Royal Society of Canada's Global Change Program. I also served as Canadian Chair of the International Joint Commission's Health Committee.

In 1990, I moved to Ottawa—Canada's capital—and established an environmental health consulting company, which provided policy services to the Canadian federal government and international agencies. This experience deepened my knowledge of environmental health and how government agencies work (or don't work).

By 2000, I had become very frustrated by government indifference and the slow pace of change on environmental health, so I decided to go back to school and do a master's degree on how to facilitate social change. This gave me an understanding of social systems that complemented my scientific knowledge of biological systems. The opportunity to combine these two ways of knowing came in 2002, when I accepted a faculty position in the Environment & Community program in the Center for Creative Change at Antioch University Seattle. I have been at Antioch ever since. Today, I teach graduate courses in environmental science, sustainability, social movements, epistemology, environmental policy, and social change. My work has been published in a variety of books, journals, and magazines.

Since moving to Seattle in 2002, I have stayed active in the environmental health movement by assembling a clearinghouse of information on health and environmental quality in Washington state and conducting a study of the economic costs of environmentally related diseases and disabilities. In the past decade, I have served on the boards or advisory groups of several U.S. environmental health organizations, including the Collaborative for Health and Environment Washington, the Institute for Children's Environmental Health, Washington Citizens for Resource Conservation, Washington Toxics Coalition, and the Sustainable Path Foundation.

THIS BOOK

So far, very little has been written about the environmental health movement. There are many books on the environmental movement, the environmental justice movement, and the science of environmental health, but only a handful on the environmental health movement. This book attempts to remedy the situation by giving the movement the recognition it so richly deserves.

Several themes run through this book. First, studying social movements, like environmental health, helps us to understand how ordinary people can bring about social change. A knowledge of history allows us to understand how people have confronted systems of power and authority in the past and puts our own struggles into context. Enabling us to see that those who fought for change in the past were often successful in the long term, history can give us hope for the future. Similarly, understanding a movement's current strategies for social change helps to design more effective strategies for tomorrow.

This type of information is in short supply because the academics who study social movements regard them as sociological and political phenomena, rather than as forces for change. Their work to explain movement inception and growth, movement grievances, participation, internal and external influences, and movement outcomes is rarely helpful for movement activists. They need a different type of information—information on why movements are successful (or not) and how to design campaigns to achieve real, lasting change. They need real-life knowledge that can be applied to real-life social problems, rather than abstract academic theories, and, although a few activist-scholars are beginning to do this type of research, more practical knowledge about social change is urgently needed.

This real-life knowledge about social change requires an interdisciplinary perspective. Looking through the lens of a single academic discipline leads to a narrow and incomplete picture. To fully understand the environmental health movement, a 360-degree view is essential—one based on all the relevant disciplines, including history, philosophy, science, economics, and politics, as well as sociology. This book uses such a perspective.

A second theme is the paradox that is central to the environmental health movement. Although anthropocentrism is a cause of many environmental health problems, the movement relies on this very worldview to make its case. By placing human health at the center of concern, the environmental health movement is anthropocentric. But the more it draws attention to the effects of the environment on human health, the more it becomes apparent that our species is not separate from or superior to the natural world. In other words, the more the environmental health movement foregrounds humankind, the clearer it becomes that we are embedded in nature. This paradox is unique to the environmental health movement.

A third theme is that environmental health is fundamentally an ethical issue, not just a matter of science and economics. It asks questions about the type of society we want and the type of world we want to bequeath to our children. Do we want a caring and compassionate society in which people's health is protected, or do we want a society that values money above everything else and sacrifices health at the altar of the Almighty Dollar? The fact that the marginalized, the poor, and minorities bear the brunt of environmental health problems reveals the fallacy of the "trickle down" effect. Environmental health, like other social movements, challenges the power-holders and the privileged.

A fourth and final theme is that achieving environmental health will take decades, if not centuries, to achieve. It is a long-term proposition that will require changing deeply engrained cultural beliefs and values. History shows that movements that transform society and achieve radical and lasting social change always take time to succeed. It took 180 years to end slavery. Women have been fighting for equality since 1848 and still haven't achieved it. The environmental health movement is only just beginning, and there is a very long road ahead.

With these themes in mind, the first part of the book explores the U.S. environmental health movement's historical and cultural roots. Chapter 1 examines the European ancestry of environmental health, and chapter 2 explains what happened when European immigrants crossed the Atlantic and created new social, political, and economic systems in the United States, based on the beliefs they brought with them. Chapter 3 picks up the story in the years after World War II and outlines how economic growth gave rise to environmentalism in the late 1960s and 1970s. The fourth chapter starts in 1978 and describes the birth of the U.S. environmental health movement and its early years.

Following this historical overview, the second part of the book examines the contemporary U.S. environmental health movement and its strategies. Chapter 5 surveys the organizations and issues that comprise it today. The movement's defining strategy—personalizing environmental issues—is analyzed in chapter 6. Chapter 7 discusses why relying exclusively on scientific arguments for environmental health is nearly always doomed to failure and considers how the movement's use of precaution avoids this fate. The next chapter describes its work on environmental justice and the right to a healthy environment. Chapter 9 considers the movement's approach to changing the economics, the markets, and business. This analysis is intended to provide insights into how the U.S. environmental health movement is advancing social change.

The book concludes with a chapter that summarizes the significance of the U.S. environmental health movement and offers suggestions about where it could go from here. Although the movement has benefitted health and is changing the way Americans think about the environment, it could do more. By learning from its own experiences and those of previous social movements, the U.S. environmental health movement could design even more effective strategies for social change.

This book focuses on the U.S. environmental health movement because this is where the international movement was born, and U.S. groups are among the oldest and most influential in the world. It's also because the need for social change is most urgent in the United States. Although this country has less than 5 percent of the world's population, it's the largest *per capita* polluter. If the United States cannot change, then there may be little hope for the rest of the world. However, this focus on the U.S. environmental health movement is not to diminish the importance of environmental health movements in other countries. On the contrary, I hope that an understanding of the situation in the United States will help environmental health movements around the world to be more successful.

In writing this book, I have become convinced that the U.S. environmental health movement is too complex to be completely encapsulated in a single volume. Its origins lie deep in the past, and it has a multifaceted history. Today, it comprises a large and unknown number of people and organizations working in nonprofit groups, communities, universities, government agencies, and other settings. Inevitably, this book omits a lot. So I hope that you will forgive me if I have failed to include the particular issue, organization, event, or person you care about.

Over the years, I have gained tremendous respect for the many people who dedicate their lives to this movement. Often working for long hours and meager salaries, they care too much to stand by and do nothing. They see the increasingly devastating effects of poor environmental quality on human health and want to do something. They want to help build a society where environmental quality actually enhances health, where no one is threatened by pollution, and where every child can grow up in a healthy, just, and sustainable world. Those of us who are part of the environmental health movement do this work because we are called to do it. Something deep inside us compels us to do it. For us, there is simply no other choice. We must act.

As the poet Adrienne Rich wrote:

> My heart is moved by all I cannot save:
> So much has been destroyed
> I have to cast my lot with those
> who age after age, perversely,
> with no extraordinary power,
> reconstitute the world.[27]

NOTES

1. The World Health Organization's definition of environmental health. Available at: http://www.who.int/topics/environmental_health/en/. Accessed August 31, 2012.
2. Caroline Webb. The mystique of the earth, an interview with Thomas Berry. *Caduceus* 59 (Spring 2003). Available at: http://www.earth-community.org/images/Caduceus %20Article_Webb.pdf. Accessed August 31 2012.
3. U.S. Environmental Protection Agency. *Report to Congress on indoor air quality: Volume 2*. EPA/400/1-89/001C. Washington, DC (1989).
4. World Wildlife Fund. *Living planet report2012*. Available at: http://awsassets.panda.org/downloads/1_lpr_2012_online_full_size_single_pages_final_120516.pdf. Accessed August 31, 2012.
5. United Nations Environment Programme. *World environment day—June 5, 2003: Water—Two billion people are dying for it!* Nairobi.
6. World Food Program and the UN Food and Agricultural Organization. *The state of food insecurity in the world 2010*. Available at: http://www.fao.org/docrep/013/i1683e/i1683e.pdf. Accessed August 31, 2012.
7. Food and Agricultural Organization of the United Nations. *Review of the state of world marine fisheries resources*. FAO Technical Paper 457. Rome (2005).
8. UN Environment Programme. *Distribution of marine litter*. Available at: http://www.unep.org/regionalseas/marinelitter/about/distribution. Accessed August 31, 2012.
9. International Union for the Conservation of Nature. *Conserving the diversity of life: About the biodiversity crisis*. Available at: http://www.iucn.org/what/tpas/biodiversity. Accessed August 31, 2012.
10. Jean-Marie Robine, S. L. Cheung, S. LeRoy, H. Van Oyen, C. Griffiths, J. P. Michel, and F. R. Hermann. Death toll exceeded 70,000 in Europe during the summer of 2003. *Les Comptes Rendus/Série Biologies* 331(2): 171–178 (2008).
11. Agence France Press (AFP), September 17, 2010.

12. World Health Organization. *Air quality and health, Factsheet No. 313*. Updated September 2011. Available at: http://www.who.int/mediacentre/factsheets/fs313/en/. Accessed August 31, 2012.

13. Kevin D. Lafferty. The ecology of climate change and infectious diseases. *Ecology* 90: 888–900 (2009).

14. Global Humanitarian Forum Geneva. *The anatomy of a silent crisis, human impact report: Climate change* (2009). Available at: http://www.ghf-ge.org/human-impact-report.pdf. Accessed August 31, 2012.

15. World Health Organization. *Preventing disease through healthy environments: Towards an estimate of the environmental burden of disease*. (2006). Available at: http://www.who.int/quantifying_ehimpacts/publications/preventingdisease/en/index.html. Accessed August 31, 2012.

16. Annette Prüss-Ustün, Carolyn Vickers, Pascal Haefliger, and Roberto Bertollini. Knowns and unknowns on burden of disease due to chemicals: A systematic review. *Environmental Health* 10(9) (2011).

17. Collaborative on Health and the Environment. *CHE toxicant and disease database*. Available at: http://www.healthandenvironment.org/tddb. Accessed August 31, 2012.

18. *EPA air quality trends*. Available at: http://www.epa.gov/airtrends/aqtrends.html. Accessed August 31 2012.

19. Carol Browner. *Statement of Carol Browner, administrator US Environmental Protection Agency*. Washington, DC: House Committee on Transportation and Infrastructure, Subcommittee on Water Resources and the Environment (October 29, 1997).

20. Charles Duhigg. Millions in US drink dirty water, records show. *New York Times*, December 7, 2009. Available at: http://www.nytimes.com/2009/12/08/business/energy-environment/08water.html?pagewanted=all. Accessed August 31, 2012.

21. Leonardo Trasande and Yinghua Liu. Reducing the staggering costs of environmental disease in children estimated at $76.6 billion in 2008. *Health Affairs* 30(5): 863–870 (2011).

22. Nicolas Z. Muller, Robert Mendelsohn, and William Nordhaus. Environmental accounting for pollution in the United States economy. *American Economic Review* 101 (5): 1649–1675 (2011).

23. Kim Knowlton, Miriam Rotkin-Ellman, Linda Geballe, Wendy Max, and Gina Solomon. Six climate change-related events in the United States accounted for about $14 Billion in lost lives and health costs. *Health Affairs* 30(11): 2167–2176 (2011).

24. Barry Commoner. *The closing circle: Nature, man, and technology*. New York: Knopf (1971).

25. Sidney Tarrow. *Power in movement: Collective action, social movements and politics*. Cambridge, UK: Cambridge University Press (1994).

26. Ken Geiser. Personal communication. June 28, 2010.

27. Adrienne Rich. *Natural resources*. In: Dream of a common language: Poems 1974–1977. New York: W. W. Norton (1978).

Part I

Historical and Cultural Roots

Chapter One

The European Ancestry of Environmental Health

Have you researched your family tree? Are you interested in where your ancestors came from? Most people are. In fact, genealogy is one of the fastest growing hobbies in the United States, replacing sewing, fishing, and even gardening in popularity. Today, millions of Americans are exploring their family history. Using wikis, social networking sites, search engines, online databases, church records, smartphone apps, and professional genealogists, they are spending more and more of their time and money to uncover their roots.

But why do we care so much about our ancestors?

We care because knowing where we came from helps us to understand ourselves. Learning about our forebears enables us to understand who we are today. Knowing where we came from gives us a sense of identity and belonging. It helps us make meaning of our lives and gives us roots that enable us to see ourselves as heirs to bloodlines that stretch back into the distant past.

Many American bloodlines originate in Europe. This isn't surprising given that between the eighteenth and twentieth centuries, tens of millions emigrated from Ireland, Germany, France, Great Britain, and other countries in southern and eastern Europe. Escaping hunger, poverty, and political unrest, European immigrants sought to build a new type of society in the United States—one based on visionary ideals and principles. Using the values they brought with them from the Old World, they created the political, economic, and social systems and institutions that define the United States today. Today, America's European ancestry can be seen in the Constitution, which is grounded in beliefs about human rights developed in France and Great Britain; our economic system, which is derived from ideas about free market economics advanced by Scotsman Adam Smith; and our justice system, which is based on English common law. Although African American, Native American, and other cultures have played a role in shaping U.S. society, these cultures were all subjugated and oppressed by the tidal wave of European culture. Because of this, there's no doubt that most of the roots of U.S. society are firmly based in Europe.

So to understand U.S. political, economic, and social systems and institutions, we need to know about their origins in Europe. This is also true for the U.S. environmental health movement and the country's environmental health problems.

The origins of environmental health go back over two and a half thousand years to ancient Greece, when philosophers, such as Plato and Aristotle, proposed that human beings were separate from and superior to the natural world. Called anthropocentrism or human exceptionalism, this idea is a cause of the ecological crisis,[1] and hence a cause of many of the environmental health problems we face today.[2] But not only does anthropocentrism cause many environmental health problems, it also defines the environmental health movement. By putting human health at the center of concern, the movement is fundamentally anthropocentric even though it's now taking account of the links between human and wildlife health.

A second, critically important idea in the development of environmental health is social justice—the belief that everyone has a right to fair and equal treatment. Originating in the European Enlightenment of the late seventeenth and eighteenth centuries, it was translated into action during the Industrial Revolution, when the appalling living and working conditions endured by millions led social reformers to call for compassion, fairness, and greater social equality. The idea of social justice has shaped concern for environmental health ever since. Other European ideas, including free market economics, are very important too, but anthropocentrism and social justice are especially significant.

This chapter describes the European ancestry of environmental health. By tracing the development of anthropocentrism, social justice, and other relevant ideas, it examines the beliefs and events that helped to shape the contemporary U.S. environmental health movement.

THE PHILOSOPHY OF ANCIENT GREECE

Even though Greece has recently come close to economic, political, and social collapse, it was once the cradle of European culture. More than two-and-a-half-thousand years ago, the philosophers of ancient Greece struggled to make sense of their lives, and in doing so they laid the foundations for Western thought. The ideas and beliefs of Socrates, Plato, Aristotle, and many, many others influence how we think and act today. The worldview we inherited from these philosophers is one in which human beings view themselves as unique. Anthropocentrism places human beings at the center of concern and asserts that our species is separate from and superior to all others, including the earth itself. This belief has allowed European culture, and other cultures affected by it, to exploit the planet's natural resources and pollute the air, water, and soil. Indeed, the anthropocentric belief that human beings are unaffected by environmental quality is often used to justify pollution. This can be seen in phrases like "It doesn't matter because it won't affect your health" or "It's buried so deep that it won't harm anyone" or "The environment will look after it."

The argument for anthropocentrism made by Aristotle (384–322 BCE) went like this: Plants can nourish and reproduce themselves, but animals have additional capacities of sensory perception and action, so animals are superior to plants. Human beings are superior to both plants and animals because in addition to sensory perception and action, they have powers of reason and inquiry. This view has dominated Western thought ever since.

For Aristotle, creating knowledge was about making generalizations based on observations of the natural world. One of the first philosophers to argue for empiricism—the idea that true knowledge is derived from sensory experience—he was a practical man who believed that the essential truth of things was defined by their physicality. In contrast, Plato (427–347 BCE), Aristotle's teacher, believed that knowledge came from reason and deduction. For him, observation revealed only a partial reality—a shadowy and incomplete view of the truth. Intellectual thought and rationalism were essential to fully grasp the essence of life. Taken together, Aristotle's empiricism and Plato's rationalism pervade Western culture to this very day. Although they can appear to be abstract concepts, they shape how we think and act in the twenty-first century.

A few ancient Greek philosophers recognized that the environment affected human well-being. Using empiricism and rationalism, they argued that good environmental conditions preserve health and, conversely, that bad ones cause illness. One of them was Hippocrates. Widely acknowledged as the father of modern medicine, he can also be credited as the father of environmental health. Written in about 400 BCE, his book *On Airs, Waters and Places*[3] asserts that water quality and a city's location affect health. The book recommends:

> Whoever wishes to investigate medicine properly, should . . . consider the qualities of the waters. . . . In the same manner, when one comes into a city to which he is a stranger, he ought to consider its situation, how it lies as to the winds and the rising of the sun; for its influence is not the same whether it lies to the north or the south, to the rising or to the setting sun. These things one ought to consider most attentively, and concerning the waters which the inhabitants use, whether they be marshy and soft, or hard, and running from elevated

> and rocky situations, and then if saltish and unfit for cooking; and the ground, whether it be naked and deficient in water, or wooded and well watered, and whether it lies in a hollow, confined situation, or is elevated and cold . . .

With these words, Hippocrates helped to transform ancient Greek concerns about health and well-being from a purely religious matter into a more earthly one. By arguing that disease was not a punishment inflicted by the gods and goddesses who lived on Mount Olympus, but rather a consequence of environmental factors, diet, and lifestyle, he created a new way of thinking. Hippocrates's ideas about the relationship between the environment and health led to the realization that people could play an active role in preventing disease and protecting their health. They were not helpless victims of vengeful deities; they had the ability to control their own destiny. This empowering belief eventually created the discipline of environmental health.

THE ENGINEERING ACHIEVEMENTS OF ROME

The engineering achievements of ancient Rome are a second historical influence in the development of anthropocentrism and environmental health. If you have ever visited Europe or northern Africa, you've probably seen them firsthand. By constructing an extensive network of roads, bridges, and dams to expand their empire, the Romans demonstrated that human beings could manipulate and control the natural world on a previously unimaginable scale. The longevity of these public works—two thousand years and counting—is a testament to their superb design and construction skills. While Greek philosophers gave birth to the idea of anthropocentrism, it was the engineers of ancient Rome who translated it into action. Centuries before steam power or electricity, their accomplishments are truly remarkable.

Although their engineering achievements demonstrate a belief in anthropocentrism, the ancient Romans also recognized that human well-being was affected by the environment. Picking up where Hippocrates left off, their empirical observations of environmental quality and its effects on health convinced them that bad air and water, sewage, debris, and a lack of personal cleanliness caused death and disease. They understood that improving environmental quality could prevent illness and protect health. So across the entire Roman Empire, they constructed massive aqueducts to provide clean water, sophisticated public latrines and drainage systems to collect sewage and other wastewater, and numerous public baths to promote personal hygiene.

Perhaps the Romans' most significant environmental health achievement was to provide their cities with safe, clean drinking water and effective drainage. According to Frontinus (ca. 40–104 CE), Rome's first water commissioner, the earliest aqueduct in the city was completed in 312 BCE. By the first century CE there were nine, with four more built shortly thereafter. Roman engineers even constructed settling basins towards the middle and end of many aqueducts to allow particulate matter, which often contained bacteria and other disease agents, to sink to the bottom. This is one of the earliest known examples of drinking water treatment.

To complement their water supply systems, many cities had complex drainage systems that ran underneath the streets and collected sewage and other wastewater. In Rome itself, there was a complex system of drains, including the "cloaca maxima"—the great sewer. It measured about ten feet wide and twelve feet high at its widest point, where it emptied into the River Tiber. Probably built during the sixth or fifth century BCE, it is still part of Rome's sewer system today.

The Romans' pragmatic approach to environmental health marked a huge leap forwards. Not only did they prevent death and disease by constructing aqueducts and drainage systems, they recognized that the health of individuals depends on measures taken at the level of the entire populace. They understood that personal actions are never enough and that collective measures are always necessary to ensure public health. Moreover, in Rome's fledgling democracy these collective measures were regarded as a government responsibility. These ideas were monumental. According to historian George Rosen, the development of public health services and their implementation by an effective government system was one of the glories of Rome.[4] Indeed, the twin beliefs that collective measures are needed to safeguard public health and that these measures are a government responsibility are central to today's environmental health movement.

THE SPREAD OF JUDEO-CHRISTIAN RELIGIONS

A third historical influence was the spread of Judeo-Christian religions, especially Christianity. Displacing the earlier pantheons of gods and goddesses and innumerable pagan spirits, this monotheistic religion dramatically expanded its reach after Roman Emperor Constantine I converted in the early fourth century CE. As Christianity spread, the Bible was adopted as the principal sacred text and the creation story in Genesis was taken as the literal truth. Not only did this story describe the origins of life, it also explained the relationship between God, man, and the natural world, saying:

> And God said, Let us make man in our image, after our likeness: and let them have dominion over the fish of the sea, and over the fowl of the air, and over the cattle, and over all the earth, and over every creeping thing that creepeth upon the earth.[5]

Historically, many have presumed that this verse sanctioned anthropocentrism because it provided man with the God-given right to exploit nature. That because man was created in God's image and was given dominion over the earth, he is its master and has unique privileges to use its resources however he wants. Today, however, many Christians disagree with this rather presumptuous and self-serving belief. They point to a later verse, Genesis 2: 15, which suggests a different relationship:

> And the Lord God took the man and put him into the garden of Eden to dress it and to keep it.

Arguing that this verse gives man unique responsibilities as well as unique rights, they claim that man's role is to act as a steward on God's behalf. As such, man is responsible for looking after the earth and all its creatures. To exploit or abuse them would be a violation of the sacred trust God placed in man. This kinder, gentler view suggests that man's relationship with nature is one of a caretaker rather than a master. St. Francis of Assisi (1181/2–1226) is often held up as a model of this way of thinking. Today, a belief in stewardship informs Judeo-Christian environmentalists. For example, the tagline of Earth Ministry, a nonprofit Christian environmental group, reads "Caring for all Creation." Similarly, the Evangelical Environmental Network says that its work is grounded in the Bible's teaching on the responsibility of God's people to "tend the garden."

Some secular commentators claim that it doesn't matter much whether man sees himself as nature's master or its steward because both views perpetuate anthropocentrism. For instance, in the late 1960s historian Lynn White claimed that Christianity's anthropocentrism was the main cause of the environmental crisis, saying, "Especially in its Western form, Christianity is the most anthropocentric religion the world has seen. . . . Man shares, in great measure, God's transcendence of nature. Christianity . . . insisted that it is God's will that man exploit nature for his proper ends."[6] Others take a more nuanced approach and argue that the idea of environmental stewardship can be used in the struggle for environmental health.

THE SCIENTIFIC REVOLUTION AND THE NATURE OF SCIENCE

A fourth historical influence in the development of anthropocentrism and environmental health is the Scientific Revolution. Although it's generally acknowledged as starting with Nicolaus Copernicus's (1473–1543) then scarcely believable notion that the sun, not the earth, was at the center of the universe, anthropocentrism became the basis for a new form of knowledge. Copernicus's heliocentric universe may appear less human-centered than Ptolemy's earlier geocentric cosmos, but the scientific approach developed by Newton, Galileo, Descartes, Bacon, and others was profoundly anthropocentric because it was based on rationality, which was deemed to be an exclusively human capacity. Indeed, the principal achievement of the Scientific Revolution was the marriage of Plato's rationalism with Aristotle's empiricism into a single coherent epistemology.[7] By combining reason and observation, it created a way of understanding that changed human consciousness forever.

The Scientific Revolution transformed how people understood human life, the natural world, and the universe. Directly challenging the power of Judeo-Christian religions, it seemed to dispense with the need for an all-powerful God. This way of knowing is now Western culture's chief epistemology. In much the same way that beliefs in religion and supernatural forces were the principal means of understanding human experience before the Scientific Revolution, so a belief in science is the principal means today.

Despite the efforts of religious fundamentalists, science is the chief explainer of life in twenty-first-century America. Just think of how we try to understand environmental health problems, or any other type of problem. We seek the facts of a situation. We identify its causes. And then we propose a solution. This approach is intrinsically scientific. Whether it's climate disruption or the latest disease pandemic, Western culture automatically looks to science and its offspring—technology—for answers, and they have responded very effectively. Funded by government agencies, corporate interests, and foundations, science and technology have significantly improved human health and well-being. For instance, immunization has helped to control many life-threatening diseases. Pasteurization and refrigeration have enhanced the safety of the food supply. And drinking water chlorination has prevented countless deaths from waterborne illnesses.

Scientific information also became a tool for developing environmental health policy. Over the past forty years, federal, state, and local governments have come to depend on it. Consider DDT and PCBs. Without scientific evidence of harm, these substances would not have been banned. Similarly, consider polluting industries. Without scientific evidence of harm, they are allowed to continue to contaminate the environment. In fact, everyone involved in environmental health policy decisions uses science in one way or another. Advocacy groups rely on it to raise public awareness and get their issues onto the policy agenda. Government agencies use it to make decisions. And corporate interests often attack it or pay for their own studies to avoid or delay government regulations. Even to the most casual observer, it's obvious that science plays a critical role in environmental health policy.

Having said this, it's important to point out that the actual role of science in environmental health policy is diminishing. Although everyone agrees that policy should be based on science, the reality is that it isn't. Because corporate interests have become so skilled at distorting, manipulating, and covering up scientific information (see chapter 7), they have undermined its role in decision making. Consider climate disruption. Despite overwhelming evidence that emissions of carbon dioxide are responsible, attempts to develop federal legislation and international agreements have run into a brick wall because of resistance from the fossil fuel industry and others who refuse to accept the science. However, despite the reality that science is becoming less important in policy decisions, there's still an illusion that it is central to the process. This explains why the U.S. environmental health movement must make its case using scientific information.

Although the Scientific Revolution marked the marriage of rationalism and empiricism, it's a lot more complicated than this. Science is also all of the qualities that characterize this union, including mechanistic thinking and beliefs in reductionism, quantification, dualism, and experimentation.

Mechanistic thinking, also known as Newtonian thinking, views the universe as a vast machine composed of small, inert particles. According to this belief, the entire physical universe, including the human body, nature, and the earth, is a finely tuned mechanism, like a clock, that operates according to precise mathematical laws. It continues ticking along, making every single aspect of life completely predictable and hence knowable. The corollary of this way of thinking is reductionism. Reductionism maintains that since everything is made up of smaller elements, the best way to understand anything is to break it apart and study the pieces. Taken together, mechanistic thinking and reductionism shape how we approach environmental health today. For instance, scientists study the health effects of one toxic chemical at a time in what is called "the substance-by-substance approach." This approach completely ignores the fact that human beings are exposed to multiple chemicals at the same time (see chapter 7).

Quantification, another quality of science, is related to mechanistic thinking and reductionism. Considered to be the epitome of pure reason, it's widely regarded as an important characteristic of true knowledge. A belief in quantification assumes that only things that can be measured are important and, conversely, that things that can't be measured aren't important. Western culture's faith in statistics and economics is an example. Politicians and policy makers rely on these quantitative disciplines to make decisions all the time. Indeed, they are considered essential.

A belief in quantification comes from early scientists who thought that the laws governing the universe were entirely mathematical. In his book *Il Saggiatore* (The Assayer) Galileo Galilei (1564–1642) declared "the universe . . . is written in the language of mathematics, and its characters are triangle, circles and other geometric figures."[8] Similarly, Rene Descartes (1596–1650) argued that mathematics is the only true form of knowledge. His work on analytical geometry, also called Cartesian geometry, helped Newton to develop calculus. Today, environmental health science relies on quantitative information. Perhaps the best example is the process of risk assessment, which is widely used by government agencies to quantify the health risks of proposed developments, exposure to toxic chemicals, and contaminated sites (see chapter 7).

Dualism, yet another quality of science, came from the idea that only human beings have the capacity for reason. Building on Aristotle's ideas, Descartes believed that thought was nonphysical, and that therefore there was a fundamental distinction between the human mind and the physical world. In effect, Descartes's disembodiment of thought created an irreconcilable gap between the mind and the body, as well as between the observer and the observed, the knower and the known, the subject and the object. Today, dualism is obvious in the way that environmental health studies are conducted. Researchers attempt to separate themselves from whatever they are studying—whether it is hazardous agents in the environment or people with an environmentally related disease. In either case, there is a clear demarcation between researchers and the objects of their studies.

Even though mechanistic thinking, reductionism, quantification, and dualism are important qualities of science, its crowning jewel is the experimental method.

The scientific method is a sequence of steps that attempts to understand reality. The first step is to identify a problem by observing a phenomenon

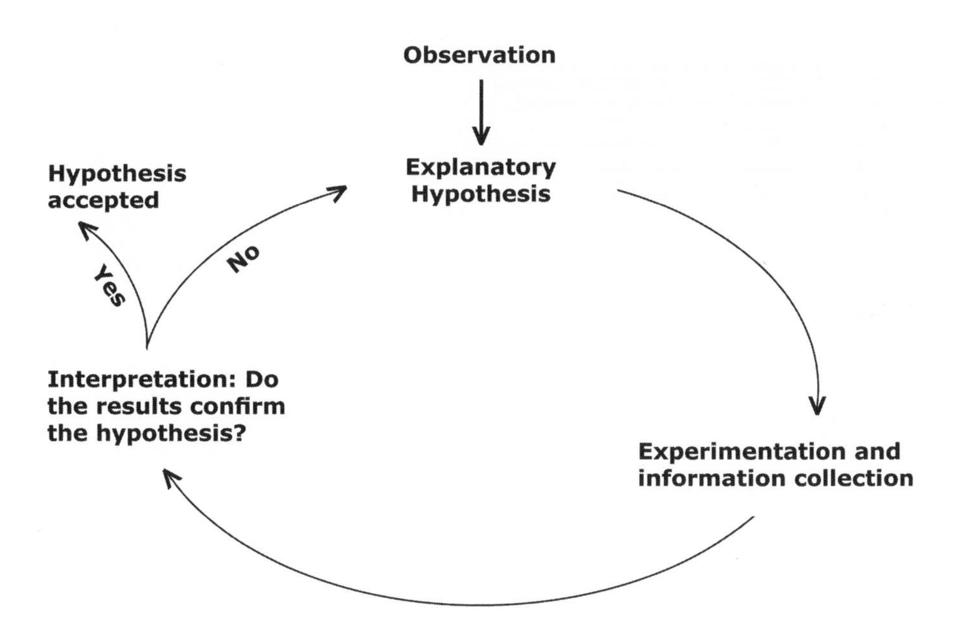

Figure 1.1. Steps in the Scientific Method

and trying to explain it in the form of a hypothesis or question. Thinking about this in terms of environmental health, you could hypothesize that Sally has cancer because of exposure to toxic chemicals, or you could ask "Did exposure to toxic chemicals cause Sally's cancer?" The second step is to test the hypothesis or question by conducting experiments and collecting information. Using the same example, you could conduct experiments on Sally's cancer or investigate her exposure to toxic chemicals. The third step is to interpret whether the information collected confirms the original hypothesis or not. If it does, then exposure to toxic chemicals may indeed be causing Sally's cancer. If it doesn't, then a new hypothesis or question is proposed to explain the observation (in this case Sally's cancer) and the process begins again. Indeed, the entire purpose of the scientific method is to explain why things happen, in other words, to try to prove causation—that something causes something else to happen. But because the scientific method can only disprove hypotheses or questions, it can never entirely prove anything, so it can never completely prove causality beyond the shadow of all doubt (see chapter 7).

Despite the fact that it's impossible to completely prove causation, science's ability to explain why things happen is very important because it allows Western culture to do two things. First, it enables us make predictions about the future, which provides a sense of security and confidence that helps us to cope with life's uncertainties. Second, understanding causation allows us to develop strategies that enable us to change the future. By helping to forecast what is likely to happen, it offers a sense of agency and control that's very reassuring. Combined, these two things help to explain why science has become Western culture's chief way of knowing.

SOCIAL JUSTICE AND THE ENLIGHTENMENT

Social justice—the belief that everyone has a right to fair and equal treatment—is a second key European idea that has influenced the environmental health movement. Implicitly anthropocentric, it's the basis for environmental justice (see chapters 4 and 8).

The idea of social justice originated during the Enlightenment of the seventeenth and eighteenth centuries. Based on the notion that every human being is entitled to equal respect and dignity, it asserts that people are entitled to human rights simply because they are human. This idea was born as Europe moved toward a worldview that was primarily scientific and secular. The new scientific way of thinking promoted critical thinking. Nothing was accepted on faith; everything was submitted to empirical-rational analysis. This examination of everything, from religion and education to war and politics, was the program of the Enlightenment philosophers, including Voltaire, Montesquieu, and Diderot. Characterized by beliefs in reason, equality, tolerance, personal liberty, freedom of expression, and democracy, this cultural and intellectual movement played a critical role in both the American and French Revolutions and influenced the French *Declaration of the Rights of Man and of the Citizen* (1789), the American *Declaration of Independence* (1776), and the *Bill of Rights* (1791).

Perhaps surprisingly, Adam Smith (1723–1790), the Scottish free market economist, was an important influence in the development of the concept of social justice. Although he is acknowledged as the father of capitalism, he was also one of the first people to present a more egalitarian view of society. In the *Wealth of Nations*,[9] originally published in 1776, Smith discussed what should be done about the "poverty problem." Instead of being contemptuous of the poor and blaming them for all criminal activity and wrongdoing, like most intellectuals of the day, he introduced the radical idea that the poor are the equals of the wealthy and that they suffer unfairly. In discussing social inequality he said: "The second duty of the sovereign, (is) that of protecting, as far as possible, every member of the society from the injustice or oppression of every other member of it. . . . Wherever there is great property, there is great inequality."[10] Although he did not offer a practical solution to the "poverty problem," Smith has been credited with preparing the way for later debates about how to advance social justice.[11]

Adam Smith's ideas about social justice influenced many others, including Tom Paine (1737–1809), the outspoken intellectual whose thinking played an important role in the American Revolution, and Jeremy Bentham (1748–1832), an English social reformer. Paine advocated the state-sponsored redistribution of wealth so that everyone could meet their basic needs. Specifically, he advocated a pension scheme to alleviate poverty among the elderly and funding it through an inheritance tax, thereby achieving the ideal of equality espoused by the Founding Fathers. Distinguishing this scheme from charity, Paine formulated the modern principle of social justice. Meanwhile, Bentham developed utilitarianism—the first comprehensive social justice theory. Based on the belief that society should be organized for the greatest benefit for the greatest number, it provided one of the policy principles that underlies European and American liberalism to this day.

But although Adam Smith, Tom Paine, Jeremy Bentham, and other Enlightenment thinkers did much to develop the notion of social justice, it was only when the Industrial Revolution aggravated inequality and unfairness to a completely intolerable level that this ideal was translated into meaningful action.

THE ENVIRONMENTAL HEALTH CONSEQUENCES OF THE INDUSTRIAL REVOLUTION

In 1819, the English poet Percy Bysshe Shelley (1792–1822) wrote:

> Hell is a city much like London—A populous and a smoky city;
> There are all sorts of people undone,
> And there is little or no fun done;
> Small justice shown, and still less pity.[12]

By then, it was virtually impossible to ignore the growing social injustice caused by the Industrial Revolution. The mass production of goods required construction of large factories to accommodate new machinery, as well as a centralized workforce to look after it. Working conditions in most factories were dreadful. Employees were often forced to work twelve to sixteen hours a day in dangerous, hot, noisy, dirty, and poorly ventilated conditions. Child labor was prevalent because children were cheaper, easier to control, and could service the machines more easily than adults—their small body size meant they could squeeze into the small spaces inside and between the machines. Industrial deaths and accidents were common.

Most peoples' living conditions were just as bad as their working conditions. The new factories required thousands of new workers, leading to a mass migration to the cities, which in turn led to a rapid increase in the demand for cheap urban housing. This demand was satisfied by constructing high density, low quality "back-to-back" houses that shared a rear wall and two side walls. Because three of the four walls were shared, these houses had few doors or windows, and the living space was very dark and poorly ventilated. In cities like Bradford, Leeds, Birmingham, and London, back-to-back housing created large, unhealthy, overcrowded slums.

To add to this, street drainage, sewage collection, and garbage pickup were nonexistent and few people had access to safe drinking water. Neighbors often shared a block of privies, which became a breeding ground for bacteria and disease. In London, garbage and raw sewage polluted the River Thames, earning the city the nickname "Venice of Drains." The first covered sewers were introduced in London in 1858, but they were so ineffectual that the windows of the Houses of Parliament were draped with sheets soaked in lime chloride to keep out the overpowering stench, and 1858 became known as the "Year of the Great Stink."[13]

Many environmentally transmitted diseases thrived in the slums of Great Britain's new industrial cities, including typhoid and tuberculosis (TB). Perhaps the most feared was cholera, nicknamed "King Cholera." This disease was especially terrifying because it was new to Great Britain in the nineteenth century. The first outbreak started in 1831, killing about twenty-three thousand people.[14] Other epidemics followed in 1848–1849, 1854, and 1867.

Like most environmentally transmitted diseases, it hit the urban poor especially hard.

But it wasn't only diseases transmitted by microorganisms that killed people. Air pollution from coal-fired steam engines and the increased use of coal as a domestic fuel caused killer fogs that lasted for days. Visual evidence can be found in Monet's art. Painted in the late nineteenth and early twentieth centuries, his pictures of London show buildings shrouded in fog and hazy urban landscapes. In fact, Monet claimed that "Without fog, London would not be a beautiful city. . . . It's the fog that gives it its marvelous depth."[15] But London "pea-soupers," as they were called, killed tens of thousands, especially in the winter. One severe episode in 1880 is thought to have killed almost twelve thousand people.[16] Others in 1882, 1891, 1892, 1948, and 1952 killed many more.

NEW POLICIES AND LEGISLATION

In response to the appalling living and working conditions and the extremely high rates of environmentally related diseases, social reformers began to call for new policies to protect environmental health. One of the first was German physician Johann Peter Frank (1745–1821). In 1779, he published the first volume of a nine-volume text called *System einer vollständigen medicinischen Polizey* (A Complete System of Medical Policy), which was completed in 1827, six years after his death. The first comprehensive treatise on environmental and public health, Frank's *System* covered a wide range of topics, including water supply and sanitation, food safety, school health, sexual health, maternal and child welfare, and the compilation of statistical records of hospitals. By arguing that government should be responsible for taking action, he became a staunch advocate for public health and measures to improve environmental quality.

In 1821, the same year that Frank died, Rudolf Virchow (1821–1902) was born. Best known as the father of modern pathology and for proposing that all cells arise from preexisting ones, he was another advocate for improving environmental quality and protecting public health. In 1848, Virchow joined forces with several other prominent German physicians to advocate for legislation based on three principles: that society, and hence government, is responsible for protecting the health of its citizens; that social and economic conditions affect health; and that government must take social and economic measures to protect health, as well as medical ones. Although Virchow and his group weren't successful,[17] the government subsequently introduced several measures based on their principles. Moreover, they still guide public health today.

Social reformers were very active in Great Britain in the nineteenth century. In 1802, concerns about child labor in textile mills led to the passage of the first *Factory Act*. Prohibiting the employment of children under nine years old and limiting their working hours, it was the first real social justice legislation ever enacted. Subsequent amendments contained further controls and authorized regular factory inspections by government-appointed officials. In 1842, the *Mines Act* (1842) was enacted in response to the findings of a Royal Commission. It excluded all women and girls, as well as boys under the age of ten, from working in underground coal mines. Even though this legislation may seem wholly inadequate by today's standards, the*Factories Acts* and the *Mines Act* were huge achievements at the time. By limiting child labor and requiring factory inspections, they helped promote the idea of social justice and the government's responsibility for advancing it.

British social reformers also targeted people's living conditions. In 1842, Edwin Chadwick (1800–1890) published the results of a study he conducted with Sir Thomas Southwood Smith (1788–1861). Called *An Inquiry into the Sanitary Conditions of the Labouring Population of Great Britain*,[18] it contained vivid descriptions of the horrendous living conditions endured by the poor. Chadwick used his report to call for public health legislation, and in 1848 he was successful. Chadwick's efforts, combined with those of an early advocacy group, called the Health of Towns Association, led to the passage of the first *Public Health Act*.

The *Public Health Act* (1848) required the creation of a new type of local government agency—Local Boards of Health—throughout Great Britain. Local Boards of Health were made responsible for upholding basic environmental health standards for water supply, sewage collection, drainage, street cleaning, public lavatories, slaughter houses, and street paving. The *Act* also established a General Board of Health to support implementation across the entire country. These measures were groundbreaking. For the first time ever, a national government guaranteed basic environmental health standards for its citizens and provided local agencies with the resources to achieve them. In this way, the *Public Health Act* of 1848 was not only a milestone in protecting public health, it was also a significant step towards social justice in Great Britain.

RECOGNIZING AND PREVENTING ENVIRONMENTALLY RELATED DISEASES

A commitment to social justice is also evident in the work of British physicians as they began to recognize and prevent environmentally related diseases. For instance, the English surgeon Percivall Pott (1714–1788) was very concerned about the health of the extremely young boys who were apprenticed to chimney sweeps. Forced up extremely narrow flues to clean them, these boys experienced burns, suffocation, respiratory disease, and eye inflammation, as well as deformed spines, kneecaps, and ankles. Pott observed that many also suffered from scrotal cancer. His 1775 report on this topic was a medical breakthrough. By noticing that chimney sweeps exposed to soot developed this potentially fatal disease, he was the first person to document that cancer could be caused by something in the environment. Chimney soot became the first identified environmental carcinogen and Pott's pioneering studies led to the *Chimney Sweepers' Act* (1788), which stated that boys could not be apprenticed as chimney sweeps before they were eight years old.

A few years later, in 1796, Edward Jenner (1749–1823), an English country doctor, was the first to show that inoculation with cowpox could make people immune to the deadly disease of smallpox. Jenner's process was rapidly adopted in many countries, and by 1801 more than one hundred thousand people had been vaccinated in England alone.[19] Mass vaccination programs were soon established in the Spanish colonies in the Americas and the Philippines, the Dutch East Indies, India, and North America. Indeed, measures to prevent smallpox were so successful that in 1979, less than two centuries after Jenner's discovery, the World Health Organization declared that smallpox had been eradicated. Because this disease was most prevalent among the urban poor, its eradication marks an important victory for social justice.

A third scientific development was the creation of new epidemiological methods to prevent environmentally transmitted diseases. In 1854, physician John Snow (1813–1858) investigated a major cholera outbreak in the Soho area of London. By talking to local residents, he found that everyone who died had drunk water from a specific pump on Broad Street. Shortly after, Snow persuaded the local council that the pump should be disabled by removing its handle.[20] Although the epidemic was likely already subsiding, his research was groundbreaking. He was the first person to use a map to show that all the cases of a disease were located close to the suspected cause, and he developed new statistical methods for studying disease. For this work, John Snow is widely considered to be the founding father of environmental epidemiology—the study of the patterns, causes, and effects of environmentally related disease.

Underlying all of these examples is the idea of prevention—that it's better to prevent illness, rather than to try to cure it after people have gotten sick. The social reformers of nineteenth-century Europe consistently advocated for prevention. But if they regarded prevention as the means to achieve health, then social justice was their goal. Taken together, these ideas led to the development of public health. Confirming the relationship between social justice and public health, public health expert Bernard Turnock said, "It is vital to recognize the social justice orientation of public health. . . . Social justice is said to be the foundation of public health."[21]

These two concepts—prevention and social justice—are also central to environmental, as well as occupational health. This is because during the European Industrial Revolution and for many years afterwards public health, environmental health, and occupational health were regarded as the same thing—what has been called environmental public health. With a goal of improving the quality of life for the masses who flocked to the cities during the Industrial Revolution, environmental public health focused on preventing disease in the urban environment. Arising from a concern for social justice, European reformers regarded public health, environmental health, and occupational health as facets of a single problem. Only later did environmental public health become separated into public health, environmental health, and occupational health (see chapter 3); But despite this separation, beliefs in prevention and social justice are central to all three. Most importantly, they are central to the U.S. environmental health movement today

NOTES

1. Early articles and books on anthropocentrism as a cause of the ecological crisis include: Lynn White Jr. The historical roots of our ecological crisis. *Science* 155(3767): 1203–1207 (1967); Arne Naess. The shallow and the deep, long-range ecology movement . *Inquiry* 16: 95–100 (1973); Morris Berman. *The reenchantment of the world.* Ithaca, NY: Cornell University Press (1981); Richard Routley. Is there a need for a new, an environmental, ethic? Proceedings of the XVth World Congress of Philosophy, *Varna* 1: 205–10 (1973). Reprinted in *Environmental Philosophy: from Animal Rights to Radical Ecology*. (ed. M. Zimmerman et al.), Englewood Cliffs, NJ: Prentice Hall (1993).

2. Benny Goodman. *Transformation for health and sustainability: Consumption is killing us*. Draft (2011). Available at: http://www.academia.edu/666114/Transformation_for_health_and_sustainability_Dualism_and_Anthropocentrism. Accessed October 31, 2012.

3. Hippocrates. *On airs, waters and places*. Part 1. Trans. by Francis Adams. Internet Classics Archive. Available at: http://classics.mit.edu/Hippocrates/airwatpl.html. Accessed on September 30, 2012.

4. George Rosen. *A history of public health*. Expanded edition. Baltimore, MD: Johns Hopkins University Press (1993). p. 25.

5. Genesis 1:26. The Bible. King James Version.

6. Lynn White Jr. The historical roots of our ecological crisis. *Science* 155(3767): 1203–1207 (1967).

7. Morris Berman. *The reenchantment of the world.* Ithaca, NY: Cornell University Press (1981).

8. Galileo Galilei. *Il saggiatore*. (1623). Trans. by S. Drake and C. D. O'Malley. In *The controversy on the comets of 1681*. Philadelphia: University of Pennsylvania Press (1960).

9. Adam Smith. *The wealth of nations: An inquiry into the nature and causes of the wealth of nations*. Indianapolis, IN: Hacket Publishing (1993). Originally published in London by W. Strahan and T. Cadell, in 1776.

10. *Op. cit.* p. 176 and 177.

11. Ben Jackson. The conceptual history of social justice. *Political Studies Review* 3: 356–373 (2005).

12. Percy Bysshe Shelley. *Hell*. Peter Bell the Third. Pt. 3 St. 1. Originally published in 1839.

13. Stephen Halliday. *The great stink of London: Sir Joseph Bazalgette and the cleansing of the Victorian metropolis*. Stroud, UK: Sutton Publishing (1999).
14. W. F. Bynum. *Science and the practice of medicine in the nineteenth century*. Cambridge, UK: Cambridge University Press (1994). p. 73.
15. As quoted in *The weasel: After the chiaroscuro smog of London, I am.* Article in the Independent Newspaper, Jan 23, 1999.
16. Health and Environment: Air Pollution. Learningspace, The Open University. Available at: http://openlearn.open.ac.uk/mod/oucontent/view.php?id=397928§ion=5.2. Accessed on September 30, 2012.
17. George Rosen. *A history of public health*. Expanded edition. Baltimore, MD: Johns Hopkins University Press (1993).
18. Report . . . from the Poor Law Commissioners on an Inquiry into the Sanitary Conditions of the Labouring Population of Great Britain. 1842. Secretary Edwin Chadwick.
19. George Rosen. *A history of public health*. Expanded edition. Baltimore, MD: Johns Hopkins University Press (1993).
20. John Snow. *The cholera near Golden Square and at Deptford.* Letter to the Medical Times and Gazette 9: 321–22. September 23, 1854. Available at: http://www.ph.ucla.edu/epi/snow/choleragoldensquare.html.Accessed on September 30, 2012.
21. Bernard Turnock. *Public health: What it is and how it works*. 4th Edition. Sudbury, MA: Jones and Bartlett (2009).

Chapter Two

Early Environmental Public Health in the United States

In the century following the American War of Independence, millions of Europeans immigrated to the United States, bringing with them their beliefs, including social justice, free market economics, and anthropocentrism. Using these and many other ideas, they created the social, economic, and political systems and institutions that define the United States today. Not surprisingly, this led to very similar environmental public health issues.

This chapter traces the history of environmental public health from the American Industrial Revolution of the early nineteenth century to the prevention of many environmentally transmitted diseases a century later. Exploring the institutions, policies, and initiatives that emerged during this time provides insights into the origins of the U.S. environmental health movement.

THE ENVIRONMENTAL HEALTH CONSEQUENCES OF THE AMERICAN INDUSTRIAL REVOLUTION

By the early nineteenth century, it was clear that the new United States needed even greater economic independence from its old colonial overlord and that this would require industrialization. Prior to independence, Great Britain regarded its American colonies as mere providers of raw materials and markets for finished goods. It kept the factories—the means for generating wealth—on British soil, thereby perpetuating its authority and control. To become truly independent, early Americans knew they had to industrialize. This led to the construction of factories, which in turn required new transportation systems, sources of energy, and a centralized workforce.

The workforce was composed of the tens of millions of immigrants who arrived from Ireland, Germany, France, and Great Britain. Hoping for a better life, most were sadly disappointed. Conditions in the cities and factories of the New World weren't much better than those they had left behind, and the same environmentally transmitted diseases killed many.

In response to outbreaks of cholera, the city of Boston established the first Board of Health in the United States in 1799, with Paul Revere (1734–1818) as its health commissioner. The new Board of Health posted signs on lampposts and held public meetings to tell people how to prevent this deadly disease.[1] This marked the beginnings of public health education in the United States.

As well as the environmentally transmitted diseases from Europe, there was a new one—yellow fever. Originating in the West Indies and transmitted by mosquitoes, yellow fever's first major recorded epidemic in the United States was in 1793 in Philadelphia. Despite quarantine measures, it soon entered New York and spread rapidly. After the disease killed thousands, New York's Common Council decided that it needed greater authority to control the spread of this deadly disease, and so in 1805, it followed Boston's example and created a Board of Health[2] to improve sanitation and environmental quality. These early concerns about environmental public health provide the origins of public health in the United States.

ENVIRONMENTAL PUBLIC HEALTH CONCERNS

The creation of Boards of Health in the new United States was significant, but social reformers wanted more, and, as in Europe, they advocated measures to improve conditions for the urban poor. One of the first champions of "sanitary science," as it was called, was John Hoskins Griscom (1809–1874), a physician and city inspector of the New York Board of Health. In 1844, he delivered a lecture on *The Sanitary Condition of the Laboring Population of New York, with Suggestions for Its Improvement*, which was published as a pamphlet in the following year.[3] Named after Chadwick's earlier report on conditions in Great Britain, Griscom's pamphlet argued that overcrowded and unhygienic housing led to illness and immorality—not the other way round, as was commonly believed at the time.

Griscom maintained that poor working people didn't have the time or money to find clean water or dispose of their wastes hygienically. Better living conditions, he claimed, would improve their health, morals, and work ethic. To achieve this, Griscom proposed that the city should ensure a safe water supply, construct a sewage system, and clean up the streets. Although his ideas weren't implemented for more than twenty years, the belief that the poor were responsible for their own health problems was too deeply embedded in society's collective consciousness for rapid social change.

In 1865, Stephen Smith (1823–1922), another physician with a social conscience, organized the first comprehensive sanitary survey of New York. Called *The Citizen's Association Report on the Sanitary Condition of the City*, the report described overflowing privies, slimy streets covered with horse manure, and slaughterhouses and fat-boiling establishments located next to overcrowded slums and tenements. It concluded:

> We, the citizens of Lower East Manhattan, declare that this city is unsuitable for human development, child development and moral development. We, citizens of all classes, have suffered from deadly diseases such as cholera, tuberculosis, small pox and pneumonia at the hands of public officials who scoff at our sufferings. We believe that housing, politics, morals and health are all intertwined and without one, we would be quite at a loss.[4]

Smith's report spurred state legislators to take action, and a year later, in 1866, they established the New York City Metropolitan Board of Health, which replaced the earlier Board of Health and created the first modern municipal public health authority in the United States. Not content to rest on his laurels, in 1872 Smith went on to cofound the American Public Health Association (APHA)—now the oldest, largest, and most diverse organization for public health professionals in the world. Established on beliefs in social justice, prevention, and science, this organization is dedicated to improving the health of all U.S. residents.

Even as environmental public health became a major concern in New York, it was also drawing attention in Massachusetts. One of its strongest advocates was Lemuel Shattuck (1793–1859), a bookseller and publisher. In his mid-forties he retired and began a new career dedicated to public service. Not only did he lay the foundations for the vital statistics systems of Massachusetts and the United States, he authored a report for the Massachusetts Sanitary Commission on environmental public health. Published in 1850, five years after John Griscom's report on conditions in New York and fourteen years before Stephen Smith's survey, Shattuck's report[5] became one of the most famous early documents on environmental public health in the United States. In particular, it emphasized the importance of environmental sanitation and preventing environmentally transmitted diseases. It also recommended that sanitary surveys should be conducted in urban communities and that local boards of health, in addition to a statewide one, should be created. Like John Griscom's pamphlet, Shattuck's report was ignored for years, although many of its recommendations were eventually implemented. In 1869, Massachusetts finally created a state board of health, followed by California, the District of Columbia, Minnesota, Virginia, Michigan, Maryland, Alabama, Wisconsin, and Illinois. Because of the work of John Hoskins

Griscom, Stephen Smith, and Lemuel Shattuck, environmental public health eventually became a consideration in the minds of state legislators.

OCCUPATIONAL HEALTH CONCERNS

As concerns about environmental public health grew, so did concerns about occupational health. In 1835, the Medical Society of the State of New York offered a prize for the best essay on the "influences of trades, professions and occupations in the United States in the production of disease," and some two years later, twenty-four-year-old Dr. Benjamin McCready (1823–1892) was declared the winner. Although McCready's essay[6] mistakenly blamed workers and their lifestyles for most occupational diseases, it was the first U.S. contribution to occupational health.

It wasn't until some thirty years after McCready's essay that state governments got involved. In 1867, Massachusetts became the first state to introduce government-authorized workplace inspections. A decade later, in 1877, it enacted the first worker safety law, requiring protective guards for the dangerous machinery used in textile mills. Others followed, and by 1900 most industrialized states had passed basic occupational safety legislation.

Despite concerns about occupational safety, little attention was given to occupational illness and the effects of hazardous substances on workers' health. Although the importance of occupational disease was recognized in Europe during the Scientific Revolution, it wasn't regarded as an important topic in the United States until Alice Hamilton's work in the early twentieth century.

Alice Hamilton (1869–1970) was an outspoken advocate for occupational health and social justice in the workplace. In 1897, she became a professor of pathology at the Woman's Medical School of Northwestern University in Chicago. There, she became increasingly alarmed about conditions in the factories and their harmful effects on workers' health. As she listened to stories about carbon monoxide poisoning in the steel mills, pneumonia and rheumatism in stockyard workers, and "phossy jaw" resulting from the use of white phosphorus in match factories, she began to realize that occupational diseases were extremely widespread in the United States.

In 1907, Hamilton began to read European studies on the subject. Contrasting European concerns with U.S. indifference, she tried to rouse her medical colleagues out of their complacency. But they were not easily persuaded and argued that occupational disease was not a problem in the United States. Clinging to the illusion that the United States was more egalitarian than Europe, they believed that working conditions in U.S. factories were better than those in Europe. Hamilton strongly disagreed. In her autobiography, she wrote:

> Everyone with whom I talked assured me that the foreign writings could not apply to American conditions, for our workmen were so much better paid, their standard of living was so much higher, and the factories they worked in so much finer in every way than the European, that they did not suffer from the evils to which the poor foreigner was subject. That sort of talk always left me skeptical. It was impossible for me to believe that conditions in Europe could be worse than they were in the Polish section of Chicago, and in many Italian and Irish tenements, or that any workshops could be worse than some of those I had seen in our foreign quarters.[7]

In 1908, Alice Hamilton published her first article on occupational disease. Later that year she was appointed to the newly formed Illinois Occupational Diseases Commission, which undertook a survey of the extent of "industrial sickness," with Hamilton as its managing director. The survey focused on "occupational poisons," specifically lead, arsenic, brass, carbon monoxide, the cyanides, and turpentine, and she took responsibility for studying the effects of lead. Soon convinced of its serious health hazards, Alice Hamilton became one of the first advocates for regulatory controls on lead and remained one for the rest of her life (lead is discussed in more detail in chapter 4).

Hamilton subsequently became assistant professor in the new Department of Industrial Medicine at Harvard Medical School—the first woman on the faculty of Harvard University—and in 1924 she was appointed to the Health Committee of the League of Nations. After her retirement in 1935, she served as medical consultant to the U.S. Division of Labor Standards. She died on September 22, 1970, at the age of 101, just three months before Congress passed the *Occupational Safety and Health Act.*

Alice Hamilton's legacy is remarkable. Most early occupational health legislation in the United States was a direct result of her work, and she was unwavering in her commitment to social justice in the workplace. Perhaps most importantly, she helped develop occupational health as a recognized scientific discipline in the United States.

THE SETTLEMENT MOVEMENT: WORKING WITH THE URBAN POOR

Alice Hamilton and many other leading female social reformers were part of the Settlement movement, which flourished from the 1880s until the 1920s. Originating in Victorian England, the movement "settled" university-educated men and women in slum areas to live and work alongside the locals. The first Settlement House was Toynbee Hall in East London, established in 1884, and the idea quickly spread across the Atlantic. The first American Settlement House—New York's University Settlement House—was founded in 1886. Other early Settlement Houses in the United States include Hull House in Chicago, started by Jane Addams and Ellen Gates Starr in 1889, as well as the Henry Street Settlement (1892) and the Lenox Hill Neighborhood House (1894), both in New York. By 1913, there were over four hundred Settlements in thirty-two states.

Located in poor, working-class areas and staffed mostly by women volunteers, Settlement Houses provided housing and food, as well as many other services. This revolutionary approach fostered a new cross-class and cross-cultural dialogue that challenged the established order and helped to advance social justice and environmental public health.

Hull House, the most famous Settlement House in the United States, offered a wide variety of facilities, including the city's first public playground, bathhouse, and gymnasium. It also had a public dispensary, a theater, a music school, a post office, libraries, art studios, a cafeteria, and meeting rooms. In addition, Hull House residents taught classes in art, bookbinding, history, literature, sewing, and other subjects; offered free concerts and lectures; and operated social clubs for children and adults. Attracting thousands of people from the surrounding poor immigrant neighborhoods, Hull House provided numerous educational and social services.

The services provided by Hull House and other Settlements were an important step forwards for environmental public health. By providing housing, food, education, and health facilities, they significantly improved living conditions and the quality of life for the urban poor. The pioneering work done by the social reformers who were part of this movement also led to legislation on child labor, occupational safety and health, public education, immigrants' rights, and pensions.

THE HOME AS AN ENVIRONMENT FOR PROTECTING HEALTH

At about the same time as the Settlement movement was gaining strength, the home came to be seen as an important setting for protecting health. For most of the nineteenth century, efforts to improve environmental public health focused on the urban environment and the workplace, with little attention given to the home. But in the 1880s, a woman called Ellen Swallow Richards (1842–1911) began to change that.

Born to a family of modest means, Ellen Swallow Richards was the first woman accepted to the Massachusetts Institute of Technology (MIT) and its first female instructor. She was also the first woman in the United States admitted to a school of science and technology and the first to earn a degree in chemistry. Over the years, Richards developed a passion for applying scientific principles to household management. Her enthusiasm blossomed into many different subjects including nutrition, sanitation, home economics, and physical fitness. Over her lifetime, she wrote many books and pamphlets including *The Chemistry of Cooking and Cleaning* (1881), *Food Materials and Their Adulterations* (1885), *Air, Water, and Food: From a Sanitary Standpoint* (1900), *Sanitation in Daily Life* (1907), and *Euthenics, the Science of Controllable Environment* (1910).[8] In 1908, she became the first president of the newly formed American Home Economics Association.

In 1892, Richards stumbled across an article written by German biologist Ernst Haeckel, proposing the new scientific discipline of "ökologie"—the study of nature's household. Her interest piqued, she wrote to him, asking permission to use the word. Haeckel replied, giving his blessing, saying he was too busy with his work in zoology.[9] Richards went on to develop the science of ecology. While working at the Lawrence Experiment Station in Massachusetts, the first environmental science—or as it was called, "sanitary chemistry"—laboratory in the United States, she conducted a large water quality study, involving more than twenty thousand samples. As a result of this study, Massachusetts established the first water quality standards in the United States and built the first modern sewage treatment plant.

Although Ellen Swallow Richards was less politically active than Alice Hamilton, she was a pioneer because she changed the way people thought. She saw the home environment, the urban environment, and the workplace environment as a continuum. Towards the end of her life, Richards began to explore a new integrative environmental science that linked them all together, calling it "euthenics"—the science of the controllable environment. Even today, Ellen Swallow Richards's ideas are remarkable. In a world where the workplace, the urban environment, and the home are treated separately, her efforts to integrate them were revolutionary.

THE PROGRESSIVE ERA AND ENVIRONMENTAL CONSERVATION

The Settlement movement and Ellen Swallow Richards's ideas about the home environment developed during a time of great political reform called the Progressive Era. Flourishing from the 1890s to the 1920s, it was based on the belief that government should help to resolve social problems, ensure economic fairness, regulate corporations, and eliminate corruption. It was led by a bipartisan coalition of Democrats and Republicans including Theodore Roosevelt, Robert LaFollette Sr., Charles Evans Hughes, and Herbert Hoover on the Republican side, and William Jennings Bryan, Woodrow Wilson, and Al Smith on the Democratic side. Such an alliance between Republicans and Democrats is scarcely imaginable in today's highly polarized political climate, but in the early twentieth century, it was a reality and led to many changes that benefitted environmental public health, including early legislation to conserve natural resources, protect consumers, and limit the power of corporations.

Some of these reforms were a result of work done by photographers, journalists, and authors who exposed injustice and corruption. Called "the muckrakers," they appealed to society's conscience. They included people like documentary photographer Jacob Riis, who revealed the dreadful conditions in urban slums in his photojournal *How the Other Half Lives* (1890);[10] journalist Ida Tarbell, who exposed the unethical tactics used by Standard Oil (now ExxonMobil) to maintain its monopoly in her book *The History of Standard Oil* (1904);[11] and author Upton Sinclair, who described the exploitation of immigrants and unsafe practices in Chicago's meatpacking industry in his novel *The Jungle* (1906).[12]

While the muckrakers exposed social injustice and corruption in the cities, the effects of industrialization on the natural environment were becoming evident. By the 1890s, America's natural resources were being aggressively exploited for private profit. Logging companies clear-cut old growth forests, mining companies excavated huge pits both on the surface and below, speculators claimed ownership of large tracts of forests and grazing land, and water resources were drained away by agricultural and industrial interests. Alarmed by this abuse, Progressive reformers passed early conservation legislation, including *The Forest Reserve Act* (1891), which resulted in the conservation of over 17.5 million acres by 1893; *The Rivers and Harbor Act* (1899), which made it illegal to throw garbage into navigable waters without a permit; and the 1903 creation of the first National Bird Preserve on Pelican Island, Florida, which marked the beginning of the National Wildlife Refuge System.

Outrage about environmental destruction also led to the birth of America's first conservation groups. In 1892, John Muir and his supporters founded the Sierra Club to, in Muir's words, "do something for wildness and make the mountains glad."[13] Muir served as the club's president until his death in 1914. In 1905, the National Association of Audubon Societies was incorporated in New York, and shortly after, the Izaak Walton League was founded in 1922 by a group of sportsmen who wanted to protect fishing for future generations. The first environmental organization with a mass membership, the League had over one hundred thousand supporters by 1925.[14]

As well as enacting conservation legislation, Progressive reformers passed measures to protect public health, including *The Pure Foods and Drugs Act* (1906), which established the Food and Drug Administration (FDA), and *The Federal Insecticide Act* (1910), which was intended to protect farmers from contaminated or mislabeled pesticides. This legislation was remarkably forward-looking and provided a framework for later legislation on environmental and public health in the 1970s (see chapter 3). The Progressive Era also saw many social reforms, including the beginning of urban planning.

THE ORIGINS OF URBAN PLANNING

Conditions in America's cities deteriorated as they grew in size and number, and by the late nineteenth century, it was obvious that something had to be done. Not only were environmentally transmitted diseases thriving in the crowded slums and tenements, so were crime, violence, prostitution, and other illegal activities. Progressive reformers believed that government had a responsibility to improve this situation.

Based on concerns about public health and well-being, early urban planning attempted to build cities that would improve people's quality of life. Not surprisingly, there were many ideas about exactly how to do this. Key among them were the ideals of the City Beautiful and Garden City movements, the visionary thinking of the Regional Planning Association of America, and New York's decision to control land use through zoning regulations.

Even though it only lasted from the early 1890s to the early 1920s, the City Beautiful movement had a significant influence on urban planning in the United States. Its proponents believed that attractive, well-designed cities would improve residents' health and quality of life and inspire a sense of civic pride that would reduce the crime rate. Hence, they proposed the construction of large, elegant public buildings, expansive parks, and wide tree-lined boulevards. By building cities that were as graceful and as culturally rich as the cities of Europe, the architects of the City Beautiful movement also hoped to attract wealthy, upper-class visitors with lots of money to spend.

One of its leaders was Daniel Burnham, director of construction for the World's Columbian Exposition, held in Chicago in 1893. He planned and built the White City, the first full-scale demonstration of the movement's ideas. The splendor of the City, the carefully considered symmetry of the buildings and its wide-open spaces were a revelation for the twenty-seven million people who visited the Exposition. Not only did the White City look magnificent, it also had state-of-the-art water, sewage, sanitation, and transportation systems and, apparently, no crime or poverty. In contrast to the real-life squalor of most American cities at the time, the White City seemed like a utopia.

The first real-life application of the City Beautiful movement's ideas was in Washington, DC. In 1901, the McMillan Commission, named after the Michigan senator who chaired the Committee on the District of Columbia, released a plan to redevelop the city. Inspired by French-born architect Pierre Charles L'Enfant's original 1791 design, McMillan's plan called for the replacement of the slums surrounding the Capitol Building with impressive neoclassical government buildings and huge public monuments. It was stunning and put Washington, DC, in the same league as Paris, London, and Rome in terms of architectural grandeur. It also deliberately attempted to link the growing power of the federal government with the vision and ideals of the country's Founding Fathers. According to historian Thomas Hines, the McMillan plan: ". . . was the first large effort to retrieve and restore the historic capital of the Founders, one of the earliest major attempts in the history of the republic to reestablish for any city a sense of continuity with its origins and with the national heritage, as expressed in architectural forms."[15]

At about the same time, the Garden City movement was advocating a very different approach. Originating in Great Britain, it called for the creation of small, compact mixed-use towns. Unlike the grandiose public buildings and stately parks of the City Beautiful movement, the Garden City movement emphasized modest housing, small-scale gardens, and local industry. The first Garden City in the United States—Radburn, New Jersey—was planned in 1929 by American architects Clarence Stein and Henry Wright. Conceived as a community that would be safe for children, Radburn was intentionally designed so that the residents wouldn't need cars. Subsequently, other Garden City communities were built, including Newport News, VA; Chatham Village, PA; Sunnyside, NY; Garden City, NY; Jackson Heights, NY; Baldwin Hills Village, CA; and the Woodbourne neighborhood of Boston, MA.

In 1923, Stein and Wright cofounded the Regional Planning Association of America (RPAA), a third influence on early land use planning in the United States. Its members were individually and collectively responsible for a wide range of visionary projects including a three-thousand-mile interstate trail system—the Appalachian Trail—and for early proposals on housing tax incentives and mortgage guarantees for poor families.[16] The RPAA also developed the idea of a regional city as an alternative to the large urban metropolis. Collaborating for a decade to plan, build, and write about the future of urban communities and regions, its forward-looking members sought to improve people's quality of life, as well as environmental public health.

Zoning was a fourth influence on urban planning. The first citywide zoning regulations in the United States were enacted in New York in 1916, in reaction to the construction of The Equitable Building, which still stands at 120 Broadway. When it was built, many people complained because it was so tall that it blocked light to neighboring buildings. These complaints led to zoning ordinances restricting what could be built where. Other cities soon followed New York's lead and enacted similar zoning regulations. However, an unintended consequence was that land values became based on their zoning designation, which became a problem for land developers.

Land developers strenuously objected to zoning because they believed they should be able to use their land anyway they wanted. In 1926, a small real estate development company in Ohio—Ambler Realty—challenged the legality of zoning all the way to the Supreme Court. Ruling that zoning is constitutional, the Court based its decision on the need to protect public health, welfare, and safety. With this decision, the Court acknowledged the connection between public health and land use and confirmed that zoning could be used to improve environmental public health. By the mid-twentieth century most U.S. cities had passed zoning ordinances that separated residential, commercial, and industrial land uses, and after World War II these ordinances helped create the American suburb (see chapter 3).

The ideas underlying the City Beautiful and Garden City movements, the RPAA, and zoning regulations helped to create urban planning, which in turn led to improvements in environmental public health. But over time, urban planning became less concerned with environmental public health. As it became a distinct discipline of its own, attention shifted from protecting people's health and improving their quality of life to the most lucrative uses of land. Increasingly falling under the influence of land speculators and developers, urban planning began to emphasize profits over people. Today, some planners are once again showing concern for health and well-being (see chapter 5).

PREVENTING ENVIRONMENTALLY TRANSMITTED DISEASES

At the beginning of the twentieth century, environmentally transmitted diseases were still the leading cause of death in the United States, but by 1930, the rates of many had been significantly reduced. As a result, fewer children died, more survived into adulthood, and overall life expectancy increased. At the same time, chronic diseases, such as cancer and cardiovascular disease, became more prevalent.

Called the demographic or epidemiologic transition, this change in the age structure of a population is seen in all developing societies. Before the demographic transition, there is a high proportion of children and fewer middle-aged and elderly people. Afterwards, the situation is reversed, with a smaller proportion of young people and a larger proportion of adults and seniors.

The causes of the demographic transition in the United States and Europe have been much debated. Physicians claim it was due to improved medical treatment. Economists claim that it was caused by the "trickle down effect" of economic growth to individuals and their families. And environmental public health professionals claim it was the result of new government-funded health and social services.

There is little evidence to support the assertion that medical advances were responsible for the reduction of environmentally transmitted diseases. Until the 1930s, there were very few treatments available, with the exception of vaccination for smallpox and diphtheria. It was only with the discovery of the first antimicrobial drugs—the sulfonamides—that physicians could cure many of these illnesses, but by then, rates had already declined significantly. Similarly, it is unlikely that the "trickle down effect" played a significant role. For most people living in poverty—the ones most affected by environmentally transmitted diseases—the "trickle down effect" was only a very slow drip until the mid-twentieth century. On the other hand, there's plenty of evidence that new government-funded health and social service programs and technologies were extremely important. Based on emerging scientific knowledge about disease causation, these programs and technologies helped to prevent many infectious diseases and improved environmental public health.

One of the new government-funded programs was the construction of public health laboratories. The first one opened in 1887 in New York, and by the 1900s, there were public health laboratories in many U.S. cities.[17] By then the germ theory of disease—the theory that infectious diseases are caused by microorganisms—had been generally accepted, and scientists were becoming knowledgeable about the microbial origins of many different illnesses and how they are transmitted through the environment. Staffed by trained scientists and engineers, public health laboratories provided a wide range of services including diagnostic testing, disease surveillance and tracking, applied research and advice on disease prevention and hygiene. They quickly became the first line of defense against environmentally transmitted and other infectious diseases. By providing services to everyone, public health laboratories made a major contribution to environmental public health.

A second major public health advance was drinking water chlorination. In 1908, Jersey City, NJ, became the first city to disinfect its drinking water with this chemical.[18] Soon after, chlorination, combined with filtration, was implemented across the entire country and mortality rates from waterborne diseases declined rapidly. This practice also transformed the function of municipal water departments, which became responsible for treating drinking water, as well as supplying it. Requiring a cadre of highly skilled professionals, drinking water treatment created a huge demand for environmental engineers and scientists. Even though the chlorination is associated with slightly higher rates of a few cancers, there is no doubt that it has prevented millions of deaths and illnesses

At about the same time, new, more hygienic ways of collecting and treating sewage were developed,[19] further reducing drinking water contamination. With the continued expansion of U.S. cities, it became evermore important to treat and dispose of sewage safely. Early treatment and disposal methods included settling basins, chemical precipitation, and irrigating land with raw sewage. Later, between 1915 and 1925, the activated sludge process was adopted in the United States and other countries.

As these and other measures helped prevent many environmentally transmitted diseases, environmental public health professionals became well-respected members of society. The American Public Health Association, established in 1872, and the Sanitarian Magazine, which started publication in 1873, played an important role in legitimizing this new profession. In the early twentieth century, creation of schools of public health at prestigious universities, such as Johns Hopkins, Columbia, Harvard, and Yale, further enhanced its status.

But even as environmental public health became a respected profession, public health changed focus and began to separate from environmental health. Although public health and environmental health were still united by a shared commitment to social justice and prevention, public health started to stress the importance of clinical care and education. For instance, New York and Baltimore introduced home visits by public health nurses, and school health clinics were set up in Boston, New York, and Rhode Island.[20] Although still housed in the same government health agencies, public health and environmental health started to grow apart, with environmental health continuing to focus on measures to improve the environment and public health emphasizing measures to benefit child and family health. To add to this, occupational health was separating from both public health and environmental health, as the work of Alice Hamilton and others created the new discipline of occupational health.

The new focus of public health and its roots in environmental public health were recognized by Charles-Edward Winslow (1877–1957), the first Chair of the Department of Public Health at Yale University. Striving to ensure that rapidly developing cities and remote rural regions were provided with sanitation, the means to control food- and waterborne diseases, and health education programs, he was a tireless champion for public and environmental health. Defining public health as "The science and the art of preventing disease, prolonging life, and promoting physical health and efficiency through organized community efforts for the sanitation of the environment, the control of community infections, the education of the individual in principles of personal hygiene, the organization of medical and nursing services for the early diagnosis and preventive treatment of disease, and the development of the social machinery which will ensure to every individual in the community a standard of living adequate for the maintenance of health,"[21] he recognized the importance of prevention, social justice, and science. This definition is still widely cited today.

This chapter has examined the evolution of early concerns about environmental public health in the United States and the social, political, and technological responses to them. Mostly dealing with the urban environment, these concerns focused on people's living and working conditions, just as they had in Europe. At the beginning of the nineteenth century, environmental public health was regarded as a single inclusive discipline in the United States, with its roots in prevention and social justice, but by the beginning of the twentieth century it was beginning to separate into three distinct, but related, disciplines—environmental health, public health, and occupational health. Moreover, these new disciplines had a basis in science. For the first time, it was possible to make a case for preventing illness and disease using scientific information, and new scientifically based programs and technologies significantly reduced the rates of many environmentally transmitted diseases. Because of this, there's no doubt that by the early twentieth century science came to play an important role in environmental health, public health, and occupational health. However, its role became even more important as increasingly sophisticated scientific studies revealed the serious environmental health problems caused by the massive economic growth that took place later in the twentieth century. As described in the next chapter, this new information contributed to the rise of U.S. environmentalism in the late 1960s and 1970s.

NOTES

1. Boston Public Health Commission. Available at: http://www.cityofboston.gov/publichealth/. Accessed September 30, 2012.
2. The New York City Department of Health and Mental Hygiene. *Protecting public health in New York city: 200 years of leadership 1805–2005* New York: New York City Department of Health and Mental Hygiene (2005).
3. John H. Griscom. *The sanitary condition of the laboring population of New York*. New York: Harper and Brothers (1845). Available on Google Books.
4. *Report of the council of hygiene and public health of the citizens' association of New York upon the sanitary condition of the city*. 2nd edition. New York: D. Appleton & Co. (1866).
5. Lemuel Shattuck. *Report of a general plan for the promotion of public and personal health, devised, prepared and recommended by the commissioners appointed under a resolve of the legislature of Massachusetts relating to a sanitary survey of the state*. (1850). Available at: http://biotech.law.lsu.edu/cphl/history/books/sr/index.htm. Accessed on September 30, 2012.
6. Benjamin McCready. On the influence of trades, professions and occupations in the United States, in the production of disease. *Transactions of the Medical Society of the State of New York* (1836/37) III. p. 146.
7. Alice Hamilton. *Exploring the dangerous trades: The autobiography of Alice Hamilton M.D.* Boston, MA: Northeastern University Press (1985, originally published in 1943).
8. *Collected Works of Ellen H. Swallow Richards* in 5 vols., edited by Kazuko Sumida. Tokyo: Edition Synapse (2007).
9. Robert Clark. *Ellen Swallow: The woman who founded ecology*. Chicago, IL: Follett Publishing Company (1973).
10. Jacob Riis. *How the other half lives: Studies among the tenements of New York*. New York: Charles Scribner & Sons (1890).
11. Ida Tarbell. *The history of Standard Oil*. McClure, Phillips and Co. (1904).

12. Upton Sinclair. *The jungle*. New York: Doubleday (1906).
13. John Muir. *The eight wilderness discovery books*. Diadem Books (1992). p. 1030.
14. John Mongillo and Bibbi Booth (eds). *Environmental activists*. Westport, CT: Greenwood Press (2001). p. 75.
15. Thomas S. Hines. *The imperial mall: The city beautiful movement and the Washington plan of 1901–02*. In *The mall in Washington, 1791–1991*. Washington, DC: National Gallery of Art (1991).
16. Kermit C. Parsons. Collaborative genius, The Regional Planning Association of America. *Journal of the American Planning Association* 60(4): 462–482 (1994).
17. George Rosen. *A history of public health*. Expanded Edition. Baltimore, MD: Johns Hopkins University Press (1993).
18. National Academy of Engineering. *Water supply and distribution history II–early years*. Available at: http://www.greatachievements.org/?id=3614. Accessed on September 30, 2012.
19. Jon Schladweiler. *Tracking down the roots of our sanitary sewers* . Available at: http://www.sewerhistory.org/chronos/roots.htm. Accessed on September 30, 2012.
20. Committee for the Study of the Future of Public Health; Division of Health Care Services. *The future of public health*. Washington, DC: National Academies Press (1988).
21. Charles-Edward. A. Winslow. The untilled fields of public health. *Modern Medicine* 2: 1–9 (1920).

Chapter Three

Environmentalism and Economic Growth

The rise of U.S. environmentalism was a consequence of the massive economic growth that took place in the United States in the twentieth century, just as concern about environmental public health was a consequence of industrialization in the eighteenth and nineteenth centuries. But unlike the earlier situation, twentieth century environmentalists could use scientific information to make their case. While the social reformers of the eighteenth and nineteenth centuries could only base their arguments on appeals for social justice, environmentalists could use science.

This chapter explores how economic growth and related events contributed to the rise of the U.S. environmental movement. By considering the growing scientific evidence that pollution affected peoples' health, the rise of the antinuclear movement, and new countercultural ideas about the environment, this chapter explains how environmentalism became a powerful political force in the United States in the late 1960s and 1970s. But first, it's important to understand how and why the economy expanded so fast in the mid-twentieth century.

POST–WORLD WAR II ECONOMIC GROWTH AND THE CREATION OF A CONSUMER SOCIETY

After the Great Depression and World War II, the U.S. economy grew extremely rapidly. Sustaining very little damage during the War years, it actually strengthened at a time when European economies were in ruins. Hence, after 1945, the United States was ideally positioned to supply the goods and services that Europe so urgently needed. The Marshall Plan (1947), named after Secretary of State George Marshall, provided the means to do this. By making available massive amounts of aid, the United States assisted in European reconstruction and became the largest supplier to the region. This created huge export markets for the United States and led to significant economic growth.

But it wasn't only new export markets that contributed to rapid economic growth. Domestic markets exploded too. A huge increase in population and the deliberate creation of a consumer society resulted in unprecedented economic expansion. Between 1946 and 1964, approximately seventy-six million babies were born in the United States, as thousands of GIs returned from the War, got married, and had children. And when the "baby boom" generation grew up, they and their parents were targeted by an increasingly sophisticated advertising industry. Promising that buying things would bring happiness and fulfillment, it offered to make the American Dream a reality for everyone.

According to Victor Lebow, an economist and retail analyst in the 1950s, "Our enormously productive economy . . . demands that we make consumption our way of life, that we convert the buying and use of goods into rituals, that we seek our spiritual satisfaction, our ego satisfaction, in consumption . . . We need things consumed, burned up, worn out, replaced, and discarded at an ever increasing rate."[1] These words reveal that the creation of a consumer society was very deliberate and that its sole purpose was to stimulate economic growth. Despite the advertising industry's altruistic promises of happiness and fulfillment, its real intention was to create a throwaway society that would require people to buy more and more "stuff."

The Marshall Plan, population growth, and the creation of a consumer society were very successful. The nation's gross national product almost tripled from about $200,000 million in 1940 to more than $500,000 million in 1960.[2] This created tens of thousands of new jobs and increased affluence, which in turn further stimulated consumption and economic growth. The expanding middle class now had enough money to buy more than just the bare necessities. Purchasing cars, televisions, radios, automobiles, homes, household appliances, clothes, and just about everything else in record quantities, consumerism revolutionized the American way of life.

But even as people were seduced into believing that consumerism was the path to happiness, many retained the socially conservative values of previous generations. In particular, women who had held jobs during World War II were encouraged to go home and become housewives again, leading to a renewed emphasis on family life.

The social pressures promoting consumerism and family life fuelled demand for single-family housing. People dreamed of owning their own homes like never before. To make this a reality and stimulate the construction industry, the federal government introduced major changes in housing policy. The

Federal Housing Act (1949) increased the availability of mortgages and the Veterans Administration provided low-cost loans to returning GIs, both with the purpose of making housing appear more affordable. At the same time, zoning regulations that were first enacted in the early twentieth century (see chapter 2) were increasingly separating industrial, commercial, and residential land uses and creating new, highly desirable areas where only residential buildings were allowed. At the same time, better public transportation, road improvements, and affordable automobiles meant that workers could live outside the city limits and commute to work. Taken together, these factors contributed to the growth of the suburbs, changing the face of the American landscape forever.

The flight of many middle-class white Americans to the suburbs brought a new set of problems, including inner-city decay, increased segregation, community fragmentation, increased smog, higher water consumption, and increasing rates of some health problems. Over the years, these issues contributed to a renewed interest in the relationship between urban planning, health, and the quality of life (see chapter 5).

There's no doubt that the unprecedented economic expansion that took place in the years following World War II had major effects on the environment and health. In fact, even before the end of World War II, the environmental health effects of pollution were becoming more obvious.

THE ENVIRONMENTAL HEALTH EFFECTS OF AIR POLLUTION

In the summer of 1943, a brown cloud formed over Los Angeles, reducing visibility and making people cough, sneeze, and complain of severe eye irritation. This was the first recorded episode of air pollution in the United States. The suspected causes were a rubber factory and backyard garbage burning. Subsequently, the factory was closed, garbage burning was banned, and Los Angeles County became one of the first in the country to hire a Director of Air Pollution Control.

Since then, there have been countless episodes of air pollution in the United States. One of the most tragic was in 1948 in Donora, Pennsylvania. In the last week of October, a thick fog descended on the town, blanketing it for almost a week.[3] Trapped by an air inversion, levels of air pollutants skyrocketed. But despite the poor visibility, residents went about their usual activities. The Halloween parade was held as usual and the local football game was played before a packed crowd, although it was difficult to see either. But many, especially the elderly and those with respiratory problems, became ill and died. By the time the fog lifted on November 1, eighteen people were dead and more than seven thousand were hospitalized or sick. In the months that followed, another fifty died. The causes of the Donora fog were never fully identified, although many suspected the local steel plant, zinc smelter, and other industrial facilities.

Affecting many industrial and urban areas, air pollution can cause severe effects on health, including bronchitis; pneumonia; lung cancer; heart disease; decreased lung capacity; eye irritation; headaches; nausea; damage to the brain, nerves, liver, and kidneys; and death. It can also aggravate asthma, emphysema, and other respiratory conditions. It's true that U.S. air quality has improved in recent years, but more than one-third of the population still lives in counties that exceed one or more federal air quality standards.[4] Moreover, tens of millions of Americans suffer from diseases associated with pollution. About twenty-seven million have heart disease,[5] twenty-five million have asthma,[6] and ten million have chronic bronchitis and emphysema.[7] And although these diseases can be caused by other factors, there's little doubt that air pollution plays a major role.

In response to worsening air quality, in 1947 California became the first state to enact air pollution legislation. The federal government didn't do anything until 1955, when it passed the *Air Pollution Control Act*. But this *Act* and its successors were largely ineffective because they left responsibility in state and local hands and did not provide these levels of government with sufficient funds to do anything. It wasn't until 1970 that public concern provided the political will to pass the *Clean Air Act*. This *Act* finally gave the federal government a strong role in controlling air pollution and contained the first national air quality standards. Although it has been amended many times since then, this *Act* is still one of this country's principal statutes for protecting environmental health.

THE ENVIRONMENTAL HEALTH EFFECTS OF WATER POLLUTION

Like air pollution, water pollution was a serious problem in the twentieth century, even though drinking water and sewage treatment had controlled many of the waterborne diseases that ravaged the United States in the nineteenth century (see chapter 2). Indeed, in 1939 President Franklin Roosevelt sent a special message to Congress emphasizing the important role the federal government could play in controlling water pollution.[8] But despite his plea, it wasn't until 1948 that Congress passed the first federal water pollution legislation. Like the later *Air Pollution Control Act* (1955), the *Water Pollution Control Act* (1948) was largely ineffective because it left primary responsibility with the lower levels of government. Although major amendments were enacted in 1956, 1961, 1966, and 1970, the *Water Pollution Control Act* (1948) wasn't completely overhauled until the *Clean Water Act* of 1972.

One of the factors that led to the passage of the *Clean Water Act* was the 1969 fire on the Cuyahoga River in Cleveland, Ohio, which was heavily contaminated with oil, trash, and other debris. Although the river had caught on fire many times before and the 1969 blaze was relatively minor, the event attracted national media attention. Time magazine declared:

> "Some River! Chocolate-brown, oily, bubbling with subsurface gases, it oozes rather than flows. "Anyone who falls into the Cuyahoga does not drown," Cleveland's citizens joke grimly. "He decays" . . . The Federal Water Pollution Control Administration dryly notes: "The lower Cuyahoga has no visible signs of life, not even low forms such as leeches and sludge worms that usually thrive on wastes. It is also—literally—a fire hazard."[9]

Public concern about water pollution was further heightened when scientists announced the "death" of Lake Erie. High levels of phosphorous from sewage and agricultural runoff led to massive algal growth, which resulted in low oxygen levels and widespread fish kills. The sight and smell of rotting fish and algae made it appear that the lake was dead. The demise of Lake Erie even made it into the first edition of *The Lorax*, the well-known children's book by Dr. Seuss:

> You're glumping the pond where the Humming-fish hummed!
> No more can they hum, for their gills are all gummed.
> So I'm sending them off. Oh, their future is dreary.
> They'll walk on their fins and get woefully weary
> In search of some water that isn't so smeary.
> I hear things are just as bad up in Lake Erie.[10]

These two events—the Cuyahoga River fire and the death of Lake Erie—contributed to the passage of the 1972 *Clean Water Act*. Like the 1970 *Clean Air Act*, this *Act* is still one of this country's principal statutes for protecting environmental health.

THE ENVIRONMENTAL HEALTH EFFECTS OF FOOD QUALITY

Food quality was another important issue. Although scientific advances such as pasteurization and refrigeration had reduced the incidence of foodborne diseases, other problems persisted and worsened. In particular, food and drugs were often adulterated, lacked labels, and were promoted with inaccurate or misleading advertising and false therapeutic claims. In 1936, Ruth deForest Lamb, the Food and Drug Administration's (FDA) chief educational officer, published a book called *The American Chamber of Horrors*,[11] which highlighted these problems using real-life examples. A year later, in 1937, public outrage reached fever pitch when more than a hundred people, including many children, died after consuming a legally marketed medicine—an “elixir of sulfanilamide”—which was actually highly poisonous. This was the straw that broke the camel’s back, and Congress soon passed the *Food, Drug and Cosmetic Act* (1938) to replace the outdated 1906 *Pure Foods and Drugs Act*. The 1938 *Act* was remarkable because it was perhaps the first federal legislation to be passed because of public concern. Led by women’s groups, citizen activists lobbied their legislators and won.

By the late 1950s, the use of synthetic pesticides, such as DDT, had increased dramatically. Although the public was not yet aware of their health hazards, a few legislators were. One of them was James J. Delaney, a Democratic congressman from New York, who chaired a House Select Committee to Investigate the Use of Chemicals in Food Products. After several oversight hearings, Delaney became convinced that there were too many pesticides and other dangerous chemicals in America’s food. So he proposed an amendment to the *Food, Drug, and Cosmetic Act*. What became known as the Delaney Clause imposed a complete ban on food additives that cause cancer regardless of any economic benefits. It stated: “the Secretary of the Food and Drug Administration shall not approve for use in food any chemical additive found to induce cancer in man, or, after tests, found to induce cancer in animals.”[12] Supporters of the Delaney Clause argued that scientists had not been able to determine a safe level for any carcinogenic chemical. This position was countered by the FDA, which argued that substances used at low levels should not be banned just because they could cause cancer at high levels. In the end, Delaney and his supporters won and this remarkably strong regulation became law in 1958.

One of the first occasions that the Delaney Clause was invoked was in November 1959 when there was a nationwide recall of cranberries suspected of contamination with the pesticide aminotriazole. Coming at the worst time of year when cranberries were in peak demand, FDA chemists worked 24/7 between Thanksgiving and Christmas testing the entire crop. Notwithstanding the ensuing negative publicity, this event convinced farmers to use pesticides more carefully.[13] Regrettably, the complete ban on carcinogens imposed by the Delaney Clause is no longer the law. In 1996, the *Food Quality Protection Act* replaced it with a “negligible risk” standard for pesticide residues.

About two years before the Delaney Clause, other health effects of contaminated food became apparent halfway around the world in Minamata, Japan. Even though this event received little publicity in the United States at the time, it became very important in the history of environmental health. In 1965, a strange new disease was first reported, affecting the nervous system and causing a lack of muscle coordination, numbness in the hands and feet, tunnel vision, and impaired speech and hearing. In extreme cases, insanity, paralysis, coma, and death followed. Minamata disease, as it was called, is a form of mercury poisoning.

From 1932 until 1968, the Chisso Corporation's chemical factory released mercury into the waters of Minamata Bay. This highly toxic metal was converted to methyl mercury by aquatic microorganisms and accumulated in local shellfish and fish, which were then caught and eaten by local residents, who subsequently developed the disease. By March 2001, 2,265 victims had been officially recognized, 1,784 had died,[14] and more than 10,000 people had received financial compensation from Chisso.[15] Lawsuits continue to this day.

Minamata disease alerted the United States and other countries to the health hazards of mercury. Although there are no officially diagnosed cases in North America, many First Nations people in Canada suspect that some of their health problems are due to high levels of methyl mercury in the fish and wild game they eat. Moreover, there are more fish consumption advisories issued for mercury in North America than for any other toxic chemical. Mercury remains a significant environmental health hazard to this day.

However, as scientists, the public, and legislators became more concerned about contaminated food and the health hazards of polluted air, water, and food in the 1940s, 1950s, and 1960s, another environmental health issue was stealing the news headlines.

THE ANTINUCLEAR MOVEMENT AND THE PRECEDENTS IT SET

The antinuclear movement was born in the years immediately following World War II. Concerns about the effects of nuclear weapons were first voiced by scientists who helped to develop the 1945 atomic bombs dropped on Hiroshima and Nagasaki. Alarmed by the scale of the suffering and damage they caused, a few of them established a journal called the *Bulletin of the Atomic Scientists* to warn the public about the dangers of nuclear weapons. Still published today, the *Bulletin* played a huge role in the development of the antinuclear movement.

By the late 1950s, the concerns of a few scientists had spread to the public and the antinuclear movement was attracting widespread support in the United States and Europe. Initially, it highlighted the health effects of radioactive fallout and the very real possibility that nuclear war would lead to the extinction of humankind. For instance, the Committee for a Sane Nuclear Policy, formed in 1957, began to advocate a ban on nuclear weapons testing. Within six years, the antinuclear movement had helped to secure agreement on the 1963 Limited Test Ban Treaty between the United States, the United Kingdom, and the Soviet Union. This treaty prohibited nuclear weapons tests in the atmosphere, outer space, and underwater. Notably, it did not ban underground nuclear tests, which continued in the United States until 1992.

In 1958, concerns about radioactive fallout prompted a major environmental health research study. In that year, biologist and antinuclear activist Barry Commoner (1917–2012) and his colleagues in the Greater St. Louis Citizens' Committee for Nuclear Information launched the Baby Tooth Survey, which collected children's baby teeth and analyzed them for strontium 90—one of the carcinogenic isotopes found in radioactive fallout. Over the next twelve years, they collected about three hundred thousand teeth. The results showed that levels of strontium 90 increased dramatically during the years of atmospheric nuclear testing and then decreased after it was banned.[16] This study is particularly notable because it was the first done by a nongovernmental organization to analyze human tissues for a hazardous substance.

The mid-1970s saw a major shift in the antinuclear movement's focus. After the energy crisis of 1973 the federal government decided to develop nuclear energy as a major source of power. Its hope was to generate cheap electricity and reduce U.S. dependence on foreign oil. Antinuclear activists were firmly opposed to this. They were supported by two prominent scientists—John Gofman and Arthur Tamplin—who, in 1969, had calculated that the allowable annual radiation exposure limit would be exceeded if a lot of new nuclear power plants were built and that even if exposure did not exceed the limit, there could be up to thirty-two thousand extra cancer deaths every year in the United States. This was the first time that a scientific risk assessment was used to support the arguments of a social movement.

Gofman and Tamplin's estimates added fuel to the fire and strengthened the antinuclear movement. Several new groups, including the Clamshell Alliance and the Abalone Alliance, were formed. The Clamshell Alliance was established by a coalition of antinuclear activists in New England in 1976. Using the same methods as civil rights protestors in the 1960s, it organized a mass sit-in at the Seabrook nuclear power plant in New Hampshire that lasted from the summer of 1976 until early 1977. Meanwhile in California, the Abalone Alliance organized activists to protest against the Diablo Canyon nuclear power plant. On August 7, 1977, fifteen hundred demonstrators rallied at the reactor's gates and forty-seven were arrested.

By the late 1970s, the antinuclear movement had broadened its attack on nuclear power and was questioning the safety of the plants themselves, as well as the lack of safe storage and disposal facilities for the highly radioactive waste they generated. Using these arguments, it gained even more public support, and, after the Three Mile Island accident in 1979, it helped stop the construction of new nuclear power plants in the United States.

In March 1979, the Three Mile Island nuclear power plant, near Harrisburg, Pennsylvania, had a partial core meltdown. The worst civilian nuclear disaster in U.S. history, it released approximately 2.4 million Curies of radioactive gases into the atmosphere. Responding to pressure from antinuclear groups and widespread public alarm, President Jimmy Carter appointed a Commission to investigate. Later that year, its report concluded "there will either be no cases of cancer or the number of cases will be so small that it will never be possible to detect them."[17] Despite this definitive statement, there are now higher than average rates of lung cancer around the Three Mile Island plant. Some speculate that they are the result of exposure to naturally occurring radon, rather than the disaster.[18] Whatever the truth, the antinuclear movement used the Three Mile Island disaster to successfully lobby for an end to the construction of new nuclear power plants in the United States.

In 1986, another disaster rallied the antinuclear movement. In April, a massive explosion and fire at the Chernobyl nuclear power plant in the Ukraine released enormous quantities of radioactive contamination into the atmosphere, which quickly spread over the western Soviet Union and Europe. Although the full extent of the health effects will never be known, there has been a huge increase in rates of thyroid cancer in children, adolescents, and young adults in the region, with almost seven thousand cases reported by 2008.[19] Higher rates of leukemia are also likely, as is long-term genetic damage. Although this disaster did not affect the United States directly, it sent chills down the spines of many Americans, who became alarmed that a similar accident could happen here.

Some twenty years later nuclear power was once again being touted as a viable option to reduce U.S. dependence on foreign oil. The *Energy Policy Act* (2005) included financial support for new nuclear plants and some states passed favorable legislation or regulations. Then, in 2012, the Nuclear Regulatory Commission approved new nuclear power reactors in Georgia—the first since 1978. Even some environmental groups, which were previously opposed to nuclear power, have softened their positions. Viewing nuclear power as the only viable option to prevent climate disruption, they now support its continued use. This has led to a split in the environmental movement. Although many activist groups, such as Greenpeace, remain adamantly opposed to nuclear power, others, such as the Environmental Defense Fund, are calling only for higher safety standards.

Meanwhile, in March 2011, yet another nuclear disaster hit. Following a major earthquake and tsunami in Japan, a series of equipment failures led to a catastrophic nuclear meltdown at the Fukushima Daiichi nuclear power plant. Large amounts of radioactive materials were released, contaminating drinking water, air, and locally grown food. The Japanese government says it will take forty years to clean up the mess. Although this disaster provided the antinuclear movement with more ammunition, it also aggravated the split in the environmental movement.

The antinuclear movement itself remains committed to its cause. This movement is critically important in the history of environmental health because it established two key precedents. First, its focus on radioactivity foreshadowed the environmental health movement's work on toxic chemicals. By opposing nuclear weapons and nuclear power, the antinuclear movement drew attention to a specific type of hazardous material—radioactive substances—setting the stage for the environmental health movement to work on another type—toxic chemicals. Second, the antinuclear movement was the first social movement to rely on scientific information to make its case. Following in its footsteps, the environmental health movement has consistently relied on science (see chapters 1 and 7).

NEW IDEAS: TOXIC CHEMICALS

As they became adults in the 1960s, the baby boom generation began to question many of the social values of their parents. It wasn't only that they were opposed to nuclear weapons; they disputed many of the basic premises of U.S. society, including beliefs about unfettered economic growth and consumerism, the role of women, drugs, racial discrimination, and conventional sexual practices. Inspired by critical theorist Herbert Marcuse (1898–1979) and a new generation of writers and intellectuals, the baby boomers created a countercultural movement that transformed U.S. society.

One of the issues explored by these writers and intellectuals was the effects of economic growth on the environment, including the consequences of increasing use of toxic chemicals, the nature of the human/environment relationship, and the link between population growth and resource depletion. Combined with growing scientific information about the health effects of pollution and the example of the antinuclear movement, their new ideas helped to inspire the U.S. environmental movement of the late 1960s and 1970s.

The most famous was Rachel Carson (1907–1964), who published her best-selling book *Silent Spring* in 1962.[20] She was already a natural history writer, but this book marked Carson's debut as a social critic. Her main thesis was that pesticide use, especially DDT, threatened wildlife and human health. Skillfully weaving together scientific fact with evocative language, she described how pesticides enter the food chain, bioaccumulate in animals and humans, and cause serious health effects. In particular, Carson made the shocking assertion that DDT and other pesticides were driving the bald eagle—the United States' national emblem—to extinction. She was right. Ten years after the publication of *Silent Spring* the federal government banned DDT and eagle populations have since recovered. Anticipating later concerns about toxic chemicals, Carson wrote:

> As the tide of chemicals born of the Industrial Age has arisen to engulf our environment, a drastic change has come about in the nature of the most serious public health problems. Only yesterday mankind lived in fear of the scourges

> of smallpox, cholera and plague that once swept nations before them. Now our major concern is no longer with the disease organisms that once were omnipresent; sanitation, better living conditions, and new drugs have given us a high degree of control over infectious disease. Today we are concerned with a different kind of hazard that lurks in our environment—a hazard that we ourselves have introduced into our world.[21]

Following publication of *Silent Spring*, Rachel Carson was viciously attacked by the U.S. chemical industry. According to one industry employee, "If man were to follow the teachings of Miss Carson, we would return to the Dark Ages, and the insects and diseases and vermin would once again inherit the earth."[22] Others criticized her scientific credentials and her character. Allegations of communism were accompanied by suggestions that she was lesbian and incapable of understanding the scientific and technical complexities of pesticides because she was a woman. Others rose to her defense. Supreme Court Justice William O. Douglas wrote a glowing review of the book, and the President's Science Advisory Committee reviewed it and completely vindicated Carson and her conclusions.

Rachel Carson died in 1964 after a long battle against breast cancer. Although she did not live long enough to see the blossoming of the environmental movement, she was an important figure in it. She was the first person to argue that synthetic chemicals pose major risks to wildlife and human health and the first person to describe the complex relationship between human health and environmental quality. These realizations make her a pioneer.

In some respects, Barry Commoner stepped into Carson's shoes. After completing the Baby Tooth Survey, he turned his attention to toxic chemicals. Using Carson's ideas as a starting point, he put toxic chemicals and their health effects in a broad socioeconomic context and argued that nothing less than sweeping social change would resolve the problem. His best-selling 1971 book, *The Closing Circle*,[23] made the case for a complete restructuring of the American economy to conform to the laws of ecology. The four laws that he proposed—everything is connected to everything else; everything must go somewhere; nature knows best; and there is no such thing as a free lunch—not only challenged beliefs in consumerism and economic growth, they also questioned Western culture's belief in anthropocentrism. Still quoted today, these laws began to change how people thought about the relationship between human beings and nature.

NEW IDEAS: DEEP ECOLOGY AND SOCIAL ECOLOGY

Barry Commoner wasn't the only person to question anthropocentrism. In 1973, Norwegian philosopher Arne Naess (1912–2009) drew a distinction between what he called the shallow and deep ecology movements.[24] Asserting that the goal of the shallow ecology movement—his name for the environmental movement—was to benefit human beings, he called for recognition of a deep ecology movement based on the principle of biocentric egalitarianism. Challenging anthropocentrism, Naess's deep ecology asserts that nature and all other species have an intrinsic right to exist and to flourish, independent of their usefulness to human beings. He called it deep ecology because unlike the shallow ecology movement, this movement attempts to change underlying cultural beliefs.

Advocated by Warwick Fox,[25] Bill Devall, and George Sessions,[26] deep ecology gave rise to environmental groups like Earth First!, the Earth Liberation Front, and the Animal Liberation Front. Regarded as radicals, their direct action tactics have angered many. But aside from their desire to inflict economic losses on the people and organizations they regard as profiting from the exploitation of the environment and animals, these organizations represent a fundamental challenge to anthropocentrism. Indeed, the presence of the Earth Liberation Front and the Animal Liberation Front on the U.S. domestic terrorist list reveals as much about the threat that these organizations pose to deep-seated cultural beliefs as it does about their actual tactics.

Social ecology provided yet another cultural critique in the 1960s. Arguing that most environmental problems stem from social problems, it claims that we can't hope to solve the ecological crisis without transforming the power relationships in society. Developed by libertarian socialist Murray Bookchin (1921–2006),[27] social ecology has had a huge influence on the environmental movement. In particular, Bookchin argued that society's power-holders, especially corporations, are the principal cause of resource depletion and pollution, not ordinary people. Asserting that corporations and power-holders exploit and manipulate the public in the same way that they exploit and manipulate the environment, he identified them as the principal cause of social and environmental problems. By highlighting social injustice and the role of corporations, social ecology laid the foundation for two of the environmental health movement's key strategies—confronting environmental injustice (chapter 8) and changing economics, the markets, and business (see chapter 9).

In the late 1980s, a vicious argument erupted between Bookchin and a few deep ecologists, most notably Dave Foreman of Earth First! Bookchin accused deep ecologists of blaming ordinary people for the ecological crisis and ignoring the corporations who were really plundering the planet. On the other hand, Foreman objected to social ecology's anthropocentrism and its disregard for the rights of other species. The acrimonious debate between them threatened to engulf the entire environmental movement. Realizing this, Bookchin and Foreman agreed to disagree. Although deep ecology and social ecology are based on very different values and assumptions, these two men realized that their shared commitment to resolving the ecological crisis was more important than their ideological differences.

NEW IDEAS: POPULATION GROWTH AND RESOURCE DEPLETION

Deep ecologists were very concerned about population growth. Indeed, one of the principles of deep ecology is that "The flourishing of human life and cultures is compatible with a substantial decrease of human population."[28] Their hope was that population control would lead to less economic development, which would decrease natural resource exploitation. They weren't the only ones concerned about population. In the 1960s and 1970s, issues of population growth, resource depletion, and environmental quality were intimately linked in America's collective psyche and in the emerging environmental movement. The nation's best-known population group, Zero Population Growth, was also an environmental group. Many of the largest environmental groups included population control as part of their advocacy platforms. As Stewart Udall wrote: "Dave Brower (then Executive Director of the Sierra Club) expressed the consensus of the environmental movement on the subject in 1966 when he said, 'We feel you don't have a conservation policy unless you have a population policy.'"[29]

In 1968, biologist Paul Ehrlich made the controversial claim that overpopulation would lead to widespread famine, mass starvation, and the deaths of hundreds of millions of people by the end of the 1970s. His book *The Population Bomb*[30] heightened public concern and contributed to demands for population control. Although history proved him wrong, Ehrlich's ideas were a forceful reminder about the link between population growth and resource depletion, echoing the concerns of nineteenth-century economist Thomas Malthus.

In 1972, four years after publication of *The Population Bomb*, the Club of Rome's report *The Limits to Growth*,[31] written by systems thinker Donella Meadows and her colleagues, drew further attention to the relationship between population growth and resource depletion. Using new computer modeling techniques to demonstrate the impossibility of unlimited economic and population growth on a finite planet, *The Limits to Growth* took the world by storm. The stark conclusion reads: "If the present growth trends in world population, industrialization, pollution, food production, and resource depletion continue unchanged, the limits to growth on this planet will be reached sometime within the next one hundred years." Like other environmental books written at the time, *The Limits to Growth* attracted public controversy. Critics claimed that the assumptions used in the computer models were invalid and that the researchers hadn't used enough data. In a *Newsweek* editorial, one Yale economist called the book "a piece of irresponsible non-sense."[32] Nevertheless, many of the trends predicted in the report have become a reality.

For many years after publication of *The Population Bomb* and *The Limits to Growth*, the environmental movement supported population control, and, in their 1985 *Environmental Agenda for the Future*, the leaders of the top ten environmental organizations in the United States dedicated a whole chapter to the subject.[33] But since the late 1990s, environmentalists have been notably silent about population control. Concerned about getting drawn into politically charged debates about reproductive rights, immigration policy, and eugenics, they have almost entirely dodged the population issue for the past fifteen years. The conscious decision to steer clear of this issue is remarkable. Even though population growth rates are declining, the world's population is still rapidly increasing, causing irreparable damage to environmental quality.

THE RISE OF ENVIRONMENTALISM

In the late 1960s and 1970s, these new ideas about toxic chemicals, deep ecology and social ecology, and population growth and resource depletion as well as mounting scientific evidence about the environmental consequences of economic and population growth roused the United States out of its complacency. Having been sleepwalking toward the American Dream since the 1950s, many woke up to the enormity of the ecological crisis for the very first time. And, seeking easy solutions, they rushed to support the growing number of environmental groups, thereby helping to create a new social movement. Many national and international groups were formed at this time, including the Environmental Defense Fund, the Natural Resources Defense Council, Greenpeace, and Friends of the Earth. Moreover, many of the conservation groups founded during the Progressive Era, including the Sierra Club, the Audubon Society, and the Izaak Walton League, gained new members.

New and old environmental groups focused their energies on protecting the environment. Tracing their origins back to the conservationists of the Progressive Era, they sought to defend nature and the natural world. In doing so, they ignored the legacy of the social reformers of the Industrial Revolution and their struggles to improve the living and working conditions of the urban poor. Instead, the new movement put on the mantle of John Muir and others who had tried to protect the American West. For many groups, the environment didn't include people or the cities where they lived and worked; protecting wilderness and wildlife was all that mattered. Despite the fact that most Americans lived in cities and towns, the young movement decided to ignore the urban environment. Indeed, in 1968, the membership of the Sierra Club voted *not* to work on urban environments.[34] This decision made the environmental movement appear elitist and discriminatory to millions of city dwellers who were concerned about pollution in their backyards and its effects on their health. It also meant that environmentalists failed to make environmental issues personally meaningful and real to many ordinary people.

Compounding this, most mainstream environmental groups were run by white, middle-class men. Growing up in affluent suburbs and educated on wealthy college campuses, they had no experience of life in the inner-city slums. They didn't know about derelict and decaying neighborhoods. They didn't know about rat- and cockroach-infested homes. They didn't know about street gangs and violence. And they didn't really want to. More concerned about resource depletion, endangered species, and the natural world, the leaders of the environmental movement ignored the concerns of poor, working-class Americans and people of color. Quite simply, the urban poor and people of color had no voice in the emerging movement.

In fact, in 1990 two regional environmental justice organizations accused the environmental movement of being racist and overly white. The Louisiana-based Gulf Coast Tenant Leadership Development Project and the New Mexico-based Southwest Organizing Project arranged for letters to be sent to many of the country's largest national groups. The letters argued that mainstream environmental groups had become isolated from poor and minority communities despite strong evidence that these communities were the chief victims of pollution. In addition, they asserted that these groups had failed to hire and promote minorities at the staff and board levels.[35] Acknowledging the validity of these arguments, Fred Krupp, CEO of the Environmental Defense Fund, one of the largest national groups in the United States, commented: "The truth is that environmental groups have done a miserable job of reaching out to minorities."[36] But although the large national groups now have a heightened awareness of environmental racism and injustice, they're still largely white. According to a study, in 2005 about one-third of all mainstream environmental organizations still had no people of color on staff.[37]

The environmental movement's decision not to embrace urban environmental health issues and its apparent racism alienated many, as did its approach to advocacy. Pioneered by the Environmental Defense Fund (EDF), which was established in 1967 in well-to-do Suffolk County on Long Island, New York, this new approach combined the skills of professional scientists and lawyers: a unprecedented alliance. For the first time, scientists and lawyers worked together to argue that DDT should be banned because it was endangering the local osprey population. This approach created a new lobbying strategy that was quickly adopted by other groups. Unlike the Progressive

Era conservation groups that relied on the persuasive powers of enthusiastic local amateurs, the new environmental groups relied on professional staff and consultants.

This new approach to lobbying was very successful. In a few short years, the environmental movement helped to pass the legislation that still provides the foundation for U.S. environmental policy today. And with their success, movement leaders began to see how a professional approach to advocacy could continue to influence environmental policy. By becoming heavily involved in litigation, rule making, and setting scientific standards, environmental lawyers and scientists sustained and even strengthened their lobbying efforts, even after legislation has been passed. But their legislative and regulatory successes came at a cost because they further distanced the young movement from ordinary Americans and reduced the importance of local volunteers.

The movement's approach to fundraising and outreach illustrates this point. Hiring paid scientists, lawyers, and other experts required large, stable sources of funding. And this, in turn, led to a new reliance on foundations, philanthropists, and direct mail campaigns. Ordinary people were only asked for their money and sometimes their signatures on petitions or prewritten advocacy letters; they were rarely invited to participate directly in setting priorities or organizing campaigns. By failing to fully engage ordinary Americans in the battle for environmental protection, the environmental movement did not capitalize on the high level of public concern about the environment. That said, there's no doubt that the young movement was 100 percent dedicated to protecting the environment. It's just that it wasn't really in touch with the public. Movement leaders sincerely believed that they could protect the environment—or rather what they considered to be the environment—by playing the Washington insiders' game. But by focusing on lobbying inside the Beltway, mainstream environmental groups failed to build strong grassroots activism on the environment.

There's little doubt that this would have been possible. The first Earth Day showed that the American public were very concerned about the environment. Founded by Senator Gaylord Nelson and held on April 22, 1970, it was the country's largest and most widespread protest to date. About twenty million people participated. National coordinator Denis Hayes helped to organize thousands of rallies in colleges, universities, cities, towns, and small communities across the United States. Protesting a variety of issues—including oil spills, nuclear power plants, polluting industries, the hazards of pesticides, wilderness loss, and wildlife extinction—demonstrators came together under a single rallying cry: Save the Earth! But movement leaders did not build on this outpouring of public concern. By failing to sustain the grassroots momentum generated by Earth Day, they squandered a valuable opportunity. Even today, with about 40 percent of Americans saying they are sympathetic to the environmental movement,[38] some national groups do little to foster local activism.

One well-known environmentalist and consumer advocate bucked this trend. At the same time as the environmental movement was professionalizing and distancing itself from the U.S. public, Ralph Nader (1934–) was breaking new ground in volunteer, grassroots activism. In 1968, he started to recruit college students from across the country to investigate government agencies responsible for protecting the environment, consumers, and workers. Dubbed "Nader's Raiders," they generated a series of reports revealing government corruption, inefficiency, and mismanagement. From this work, a network of college-based organizations emerged called Public Interest Research Groups (PIRGs). Over the years, PIRGs have addressed a variety of environmental health issues including toxic contamination in and around Love Canal, New York, and nuclear power plants in California. Today, there are still PIRGs in about twenty states and Canada.

Not only did the young environmental movement fail to sustain and grow grassroots activism and ignore concerns about urban pollution and health, it also failed to connect with the public health institutions that had been protecting the environment for more than a century (see chapter 2). Because environmental groups traced their lineage from conservation groups, they did not understand the role of public health and its links with the environment. In doing so, the environmental movement ignored a powerful set of allies who could have helped to diversify its base of support, advance its political agenda, and strengthen its cause.

THE EPA AND THE FINAL SEPARATION OF PUBLIC AND ENVIRONMENTAL HEALTH

Environmentalists weren't the only ones who ignored the obvious links between public health and environmental health. The federal government did exactly the same thing. Even though responsibility for environmental health had been located in the Public Health Service for decades, in 1970 Republican President Richard Nixon established the Environmental Protection Agency (EPA) as a new and entirely separate government agency. By removing responsibility for environmental health from the Public Health Service, he institutionalized the final separation between public health and environmental health that had begun in the early twentieth century (see chapter 2). Although creating a new agency shone a spotlight on the federal government's commitment to environmental protection, it resulted in an awkward, unnecessary, and costly fragmentation of government responsibilities.

In retrospect, it would have made more sense to retain responsibility for environmental health in the Public Health Service and expand its authority.

Instead, Nixon chose to create a new agency which consolidated a variety of federal research, monitoring, standard-setting, and enforcement programs. In particular, federal authority for water pollution, air pollution, pesticides, and some aspects of food and radiological safety were transferred to the EPA. Because these programs were staffed by engineers and scientists, the new agency naturally focused on engineering and technical solutions. Public health's emphasis on social justice and prevention was largely forgotten in favor of technology-based "end-of-pipe" solutions. Unfortunately, many states followed the lead of the federal government and set up new environmental departments that used a similar approach and were separate from existing public health agencies.

By the late 1980s, the problems of creating environmental agencies and separating jurisdictional authority for public health and environmental health were becoming clear to everyone, even the medical profession. In 1988, the Institute of Medicine declared "The removal of environmental health authority from public health has led to fragmented responsibility, lack of coordination, and inadequate attention to the public health dimensions of environmental health issues."[39] This fragmentation is still a problem today. In 2009, the Agency for Toxic Substances and Disease Registry (ATSDR) and the Centers for Disease Control and Prevention (CDC) launched a *National Conversation on Public Health and Chemical Exposure* to address this separation and strengthen the public health role in managing toxic chemicals. Released in 2011, its Action Agenda[40] included recommendations for more scientific information and legislative reform. Although these recommendations were developed by hundreds of participants from different sectors during countless hours of discussion, the original goal of the *National Conversation*—to bridge the institutional chasm between environmental and public health in the United States—has proven elusive.

In countries where environmental health is still a responsibility of public health agencies, such as the United Kingdom and Canada, this chasm is narrower or nonexistent. Public health professionals work alongside their environmental health colleagues to plan, design, implement, and review policies and programs together. Because they're all part of the same agencies, there is better collaboration and coordination, and less competition for resources. Moreover, the traditional public health values of preventing disease and social justice automatically permeate environmental health programs in a way that is often missing in the United States.

THE RELATIONSHIP BETWEEN THE ENVIRONMENTAL MOVEMENT AND THE LABOR MOVEMENT

As well as institutionalizing the separation between public and environmental health, the federal government also formalized the separation between occupational health and environmental health. For decades, federal responsibility for occupational safety and health rested with the Office of Industrial Hygiene and Sanitation, also part of the Public Health Service. This meant that programs on occupational health and environmental health were colocated with public health. But in 1970—the same year that the EPA was established—President Nixon created another new agency—the Occupational Safety and Health Administration (OSHA).

By this time, thanks to Alice Hamilton and others, the labor movement had become concerned about occupational safety and health and had successfully lobbied for federal legislation, including the *Metal and Nonmetallic and Mine Safety Act* (1966), the *Coal Mine Safety and Health Act* (1969), and the *Contract Work Hours and Safety Act* (1969). In 1970, this legislation was followed by passage of the *Occupational Safety and Health Act*, which established OSHA.

Few environmentalists lobbied for the *Occupational Safety and Health Act*. Preoccupied with the natural environment, they failed to see the obvious links between occupational health and environmental health. Exceptions included Friends of the Earth, a group called Environmental Action, and Ralph Nader. Together with his young activist researchers ("Nader's Raiders"), Nader lobbied hard for the *Act's* health and safety provisions. He understood the importance of creating a strong, long-term partnership between the environmental movement and the labor movement and tried to use health to build a bridge between them. In the late 1960s and early 1970s, his public interest research groups (PIRGs) and Health Research Group investigated both occupational and environmental health issues and encouraged grassroots activism. Meanwhile, on the labor side, Anthony Mazzocchi of the Oil, Chemical and Atomic Workers worked hard to forge alliances between local environmentalists and the growing occupational health movement of the early 1970s.

In 1972, Mazzocchi's efforts began to pay off. In that year, the first Committee on Occupational Safety and Health (COSH) was formed—the Chicago Area Committee for Occupational Safety and Health (CACOSH). Others soon followed, including the Massachusetts Coalition for Occupational Safety and Health (MassCOSH), the Philadelphia Area Project on Occupational Safety and Health (Philaposh), and the Pittsburgh Area Committee for Occupational Safety and Health (PACOSH). In all, about forty COSHs were established in the United States. Their members included workers, physicians, lawyers, industrial hygienists, scientists, environmentalists, community organizers, and journalists. Realizing that the workplace was a major source of exposure to toxic chemicals for employees and community residents alike, COSHs coordinated local opposition to pollution inside and outside the factory walls. Their efforts to bridge the gap between occupational and environmental health were remarkable.

Not only did most environmentalists fail to appreciate the links between occupational health and environmental health, they didn't trust the labor movement. Suspicious of labor's support for economic growth, they accused union officials of ignoring its environmental consequences. On the other hand, union officials were wary of the middle-class composition of many national environmental groups and their apparent disinterest in wages, job security, and other labor issues.

Despite this gulf, there were a few short-term efforts at mutual support. In 1973, three years after passage of the *Occupational Safety and Health Act*, the Sierra Club, the Wilderness Society, and the National Parks and Conservation Association supported workers at Shell Oil who went on strike for health and safety protections.[41] Then in 1976, biologist Barry Commoner and economist Hazel Henderson helped to form an organization called Environmentalists for Full Employment (EFFE). Commoner and Henderson had already begun to counter industry arguments that environmental protection costs jobs by pointing out that the opposite was likely to be true. EFFE took their work further and made the case that environmental protection would actually create jobs, as well as improve worker and community health. In later years, other alliances were formed, including the campaigns for the right to know (see chapter 4) and the BlueGreen Alliance (see chapter 5).

THE *TOXIC SUBSTANCES CONTROL ACT* AND OTHER ENVIRONMENTAL LEGISLATION OF THE 1970S

In the 1970s, Congress passed at least twelve major laws to protect the environment, making the United States a world leader in environmental policy. Years ahead of others, this country became a model for the international community. By then, it was obvious that leaving responsibility for environmental protection to state and local governments didn't work. Not only was environmental quality deteriorating across the country, but there was an uneven national patchwork of legislation and regulations. Moreover, most state and local programs were starving for resources and enforcement was inadequate or nonexistent.

The environmental movement took advantage of this situation to lobby for strong federal intervention in the form of a national framework that would provide everyone with the same basic level of environmental protection. Using its new professional approach to lobbying, it went to work in

Congressional hearing rooms and corridors. And it was very successful. Almost all of America's foundational environmental legislation was passed in less than a decade, including the *National Environmental Policy Act* (1969), the *Clean Air Act* (1970), the *Clean Water Act* (1972), the *Federal Environmental Pesticides Control Act* (1972), the *Safe Drinking Water Act* (1974), the *Toxic Substances Control Act* (1976), and the *Resource Conservation and Recovery Act* (1976). This legislation is summarized in the table 3.1

One of the main pieces of federal environmental legislation passed in the 1970s was the *Toxic Substances Control Act* (1976). TSCA was intended to provide the EPA with the regulatory tools necessary to protect the public from the environmental health risks of toxic chemicals and to ensure that there is enough safety information available. Unfortunately, it has not lived up to either of these goals.

In many ways, TSCA was doomed from the start. By "grandfathering" nearly all the chemicals that were used in 1976 (about sixty-two thousand) and excluding them from any review or testing requirements, the *Act* created a monumental loophole for the chemical industry. The manufacturers of existing chemicals didn't have to do a thing. It's true that manufacturers of new substances are required to submit a premanufacturing notification and that the EPA can regulate new substances if it finds they pose an "unreasonable risk to human health or the environment," but the EPA must first prove that a substance is harmful. In other words, there is an *a priori* assumption that new chemicals are safe, until the EPA proves they are harmful. This permissive approach runs counter to the idea of prevention. A chemicals policy based on preventing environmentally related diseases and disabilities would require chemical companies to show that substances are safe before they could manufacture them. Even worse, it's virtually impossible for the EPA to prove that chemicals pose an "unreasonable risk" because of the extremely high scientific standard of proof required (see chapter 7). In fact, since 1976, only five substances have been banned under TSCA, including PCBs, halogenated chlorofluoroalkanes, dioxins, and hexavalent chromium. That's five out of more than eighty thousand now in use in North America. Not a single chemical has been banned in the past twenty-odd years. Even efforts to use TSCA to ban asbestos—a known carcinogen—failed.

TSCA's failings are well-known and have been discussed for years by academics[42] and environmentalists.[43] Even the federal government has acknowledged its shortcomings.[44, 45] But despite widespread agreement about the need for reform, very little has actually happened. In fact, TSCA has the unenviable distinction of being the only major piece of environmental legislation passed in the 1970s that has not been significantly amended. All of the others have been revised. But despite these amendments, the United States has long since lost its role as a world leader in environmental protection. Today, this distinction belongs to Europe, where initiatives, such as the

Table 3.1. Major Federal Environmental Legislation 1969–1977

Legislation	Year	Main Points
National Environmental Policy Act	1969	Required federal agencies to consider the environmental impacts of their proposed actions and reasonable alternatives to those actions. Created the Council on Environmental Quality and established a national environmental policy.
Occupational Safety and Health Act	1970	Created the Occupational Safety and Health Administration (OSHA) and made it responsible for setting standards and conducting inspections to ensure that employers provide safe and healthful workplaces.
Clean Air Act Amendments	1970	Required the EPA to set national air quality standards and certain emission limits. Required states to develop implementation plans by specific dates. Required reductions in automobile emissions.
Consumer Product Safety Act	1972	Established the Consumer Product Safety Commission (CPSC). Authorized the CPSC to set standards when it finds an unreasonable risk of injury associated with a consumer product. Authorized to ban a product if there is no feasible standard.
Clean Water Act Amendments	1972	Contained national objectives to restore and maintain the chemical, physical, and biological integrity of the nation's waters. Set up the pollutant discharge permit system for point sources. Increased federal grants to states for the construction of sewage treatment plants.
Federal Environmental Pesticides Control Act (amended the 1947 *Federal Insecticide, Fungicide and Rodenticide Act*)	1972	Required registration of all pesticides sold, distributed, or applied in the United States. Authorized the EPA to issue experimental use permits, review registered pesticides, and to cancel or suspend registration under specified conditions.
Lead-Based Paint Poisoning Prevention Act	1973	Authorized a ban on lead-based paint in cooking, drinking, and eating utensils; homes built or remodeled by the federal government; and in children's toys and furniture.

Safe Drinking Water Act	1974	Authorized the EPA to set national health-based standards for drinking water to protect against both naturally occurring and man-made contaminants that may be found in drinking water.
Toxic Substances Control Act	1976	Authorized premarket testing of chemicals. Allowed the EPA to ban or regulate the manufacture, sale, or use of any chemical that poses an unreasonable risk to health or to the environment. Banned most uses of PCBs.
Resource Conservation and Recovery Act	1976	Authorized the EPA to control hazardous waste from "cradle to grave," including the generation, transportation, treatment, storage, and disposal of hazardous waste. Established a framework for the management of nonhazardous solid wastes.
Clean Air Act Amendments	1977	Contained requirements for sources in areas not attaining the National Ambient Air Quality Standards ("non-attainment areas"). Set new standards to prevent significant deterioration in clean air areas. Postponed deadlines for compliance with air quality standards and auto emissions.
Clean Water Act Amendments	1977	Extended deadlines for industries and municipalities to meet treatment standards. Set national standards for industrial pretreatment of wastes. Increased funding for sewage treatment plant construction grants.

REACH chemical management system, are superior to anything in the United States (see chapter 5).

This chapter has explored how the massive economic growth that took place in the United States in the twentieth century resulted in environmental health effects and how these effects, combined with the antinuclear movement and new ideas about the environment, gave rise to environmentalism. This chapter has also criticized the young environmental movement for failing to address the urban environmental health issues of poor, working-class Americans and people of color and for failing to support local, grassroots activism. But despite these shortcomings, the environmental movement succeeded in making the United States a world leader in environmental protection legislation in the 1970s, even if some of that legislation, like TSCA, is seriously flawed. The next chapter describes how local activists took advantage of the environmental movement's shortcomings and gave birth to a grassroots, urban environmental health movement.

NOTES

1. Victor Lebow. Price competition in 1955. *Journal of Retailing* 31(1) Spring (1955).
2. Christopher Conte, Albert R. Carr, George Clark, and Kathleen E. Hug. *Outline of the US economy*. Washington, DC: US Department of State, Office of International Information Programs (2001).
3. Devra Davis. *When smoke ran like water: Tales of environmental deception and the battle against pollution*. New York: Basic Books (2002).
4. U.S. EPA. Air quality trends. Available at: http://www.epa.gov/airtrends/aqtrends.html. Accessed September 30, 2012.
5. Veronique L. Roger, Alan S. Go, Donald M. Lloyd-Jones, et al. Heart disease and stroke statistics —2012 update: A report from the American Heart Association. *Circulation*. 125: e2–e220 (2012).
6. Centers for Disease Control and Prevention. *Asthma in the US*. Available at: http:// www.cdc.gov/vitalsigns/asthma/. Accessed October 31, 2012.
7. American Lung Association. *Trends in COPD (Chronic Bronchitis and Emphysema): Morbidity and Mortality*. (August 2011).
8. William Andreen. The evolution of water pollution control in the United States—state, local, and federal efforts, 1789–1972, Part II. *Stanford Environmental Law Journal* 22: 215–294 (2003).
9. America's sewage system and the price of optimism. *Time Magazine*. Friday, August 1, 1969. Available at: http://www.time.com/time/magazine/article/0,9171,901182-1,00.html. Accessed September 30, 2012.
10. Dr. Seuss and Theodor Seuss Geisel. *The lorax*. New York: Random House (1971).
11. Ruth deForest Lamb. *The American chamber of horrors*. New York: Farrar and Rinehart (1936).
12. *Food, Drug and Cosmetic Act*, Food Additives Amendment, also known as the Delaney Clause, 1958 PL 85-929.
13. U.S. Food and Drug Administration. *The story of the laws behind the labels*. Part II 1938—The Federal Food, Drug, and Cosmetic Act. FDA Consumer. June 1981.
14. Ministry of the Environment, Government of Japan. *Minamata disease the history and measures. 2: Outbreak of Minamata disease*. Available at: http://www.env.go.jp/en/chemi/hs/ minamata2002/ch2.html. Accessed on September 30, 2012.
15. National Institute for Minamata Disease. *Minamata disease archives*. Available at: http:/ /www.nimd.go.jp/archives/english/index.html. Accessed on September 30, 2012.
16. The St. Louis baby tooth survey. Washington University School of Dental Medicine, legacy of achievement. Available at: http://beckerexhibits.wustl.edu/dental/articles/babytooth.html. Accessed on September 30, 2012.
17. *Report of the President's commission on the accident at Three Mile Island* (1979). Available at: http://www.pddoc.com/tmi2/kemeny/index.htm.Accessed September 30, 2012.
18. R. William Field. Three Mile Island epidemiologic radiation dose assessment revisited: 25 years after the accident. *Radiation Protection Dosimetry* 113(2): 214–217 (2005).
19. UNSCEAR. *Sources and effects of ionizing radiation*. Volume II, Scientific Annexes C, D, and E. New York: United Nations (2011).
20. Rachel Carson. *Our silent spring*. New York: Crest Books. Originally published by Houghton Mifflin (1962).
21. Rachel Carson. *Our silent spring*. New York: Crest Books. Originally published by Houghton Mifflin (1962).
22. As cited in Linda Lear. *Rachel Carson: Witness for nature*. New York, NY: Henry Holt (1997).
23. Barry Commoner. *The closing circle: Man, nature and technology*. New York, NY: Alfred Knopf (1971).
24. Arne Naess. The shallow and deep, long-range ecology movement: A summary. *Inquiry* 16: 95–100 (1973).
25. Warwick Fox. *toward a transpersonal ecology: Developing new foundations for environmentalism*. Albany, NY: SUNY Press (1995).

26. Bill Devall and George Sessions. *Deep ecology: Living as if nature mattered.* Layton, UT: Gibbs Smith (1985).
27. Murray Bookchin. *The ecology of freedom: The emergence and dissolution of hierarchy.* Cheshire Books (1982).
28. Bill Devall and George Sessions. *Deep Ecology* . Gibbs M. Smith (1985) p. 85–88.
29. Steward L. Udall. *The quiet crisis and the next generation.* Salt Lake City: Peregrine Smith Books (1988). p. 239.
30. Paul Ehrlich. *The population bomb.* New York: Sierra Club-Ballantine Books (1968).
31. Donella Meadows, Dennis Meadows, Jørgen Randers, and William W. Behrens III. *The limits to growth: A report for the Club of Rome on the predicament of mankind.* London, UK: Earth Island (1972).
32. Henry Wallich. *Newsweek* editorial. Published on March 13, 1972.
33. John Adams, Louise Dunlap, Jay Hair, Frederick Krupp, Jack Lorenz, Michael McCloskey, Russell Peterson, Paul Pritchard, William Turnage, and Karl Wendelowski. *An environmental agenda for the future.* Washington, DC: Island Press (1985).
34. Peter Montague. The environmental movement: Part 6 changing the climate of opinion. *Rachel's News.* Issue 746, March 13, 2002. Available at: http://www.rachel.org/?q=en/node/5508. Accessed September 30, 2012.
35. Robert Gottlieb. *Forcing the spring: The transformation of the American environmental movement.* Washington, DC: Island Press (1993).
36. As quoted in Robert Gottlieb. *Forcing the spring: The transformation of the American environmental movement.* Washington, DC: Island Press (1993). p. 261.
37. Dorceta Taylor. *Diversity in environmental institutions: Summary results of the MELDI studies.* Ann Arbor, MI: University of Michigan School of Natural Resources and Environment (2005).
38. Gallup. Environment. Available at: http://www.gallup.com/poll/1615/environment.aspx. Accessed September 30, 2012.
39. Institute of Medicine. *The future of public health.* Washington, DC (1988).
40. *Addressing public health and chemical exposures: An action agenda.* Available at: http://www.nationalconversation.us. Accessed on September 30, 2012.
41. Deborah Shapely. Shell strike: Ecologists refine relations with labor. *Science* 180(4082): 166 (1973).
42. See for example, Lowell Center for Sustainable Production, University of Massachusetts at Lowell. US federal chemicals policy, *Toxic Substances Control Act.* Available at: http://www.chemicalspolicy.org/chemicalspolicy.us.federal.tscaindetail.php. Accessed on September 30, 2012.
43. See for example, Joseph H. Guth, Introduction to the *Toxic Substances Control Act*, Science and Environmental Health Network (2006). Available at: http://sehn.org/lawpdf/TSCASummary.pdf. Accessed on September 30, 2012.
44. General Accounting Office. *Toxic Substances Control Act:* Legislative changes could make the *Act* more effective. GAO/RCED 94-103 September 26, 1994.
45. General Accounting Office. *Chemical regulation: Options exist to improve EPA's ability to assess health risks and manage its chemical review program.* GAO 05-458 (June 2005). Available at: http://www.gao.gov/new.items/d05458.pdf. Accessed on September 30, 2012.

Chapter Four

The Birth of the U.S. Environmental Health Movement

In 1978, a young stay-at-home mom named Lois Gibbs first raised the alarm about the health effects of toxic chemicals leaking from an abandoned waste site in Love Canal, New York. Gibbs and her neighbors faced the scary realization that their children were seriously ill just because of where they lived. They fought the government and won. The disaster at Love Canal marks the birth of the environmental health movement. Although Rachel Carson's 1962 book *Silent Spring* alerted the public to the health effects of toxics some sixteen year earlier, it wasn't until 1978 that local activists began to express their concerns and organize collective action. From these humble beginnings, environmental health has grown into the national and international movement it is today.

In giving birth to the environmental health movement, Lois Gibbs and other local activists took advantage of the shortcomings of the environmental movement and created a new social movement based on community concerns. Feeling ignored by national environmental groups, their professional advocacy approach and their focus on federal legislation, citizen activists began to speak up for themselves. In particular, they laid claim to the urban environmental health issues ignored by environmentalists. Following in the footsteps of the social reformers of the Industrial Revolution, they took action to improve environmental conditions in America's towns and cities.

This chapter examines the birth of the U.S. environmental health movement. By considering the key issues that shaped its development, this chapter presents an overview of the movement's early years from its inception in 1978 to approximately 2000.

LOVE CANAL AND ITS AFTERMATH

The tragedy at Love Canal began to unfold in the spring of 1978.[1, 2, 3] Lois Gibbs, president of the Love Canal Homeowners' Association, and her neighbors discovered a toxic dump containing twenty thousand tons of chemical wastes underneath the local school and their homes. Becoming concerned that her children's recurring health problems were caused by this pollution, Gibbs organized a petition to pressure the local School Board to investigate the problem and close the school, if necessary. After knocking on a few doors, she quickly realized that the community's health problems were much worse than she originally thought. And after she had knocked on a few more, the petition began to attract media interest. In her own words: "Men, women, and children suffered from many conditions—cancer, miscarriages, stillbirths, birth defects, and urinary tract diseases. The petition drive generated news coverage and helped residents come to the realization that a serious problem existed."[4]

The origins of the toxic wastes at Love Canal go back to the 1890s. The Canal was named after William T. Love, who envisioned a channel between the Upper and Lower Niagara Rivers to generate power for the area. Unfortunately for Love, his plan was never realized. He had only just started digging before his money ran out. After the project was abandoned, the canal gradually filled with water, and in the 1920s it was used as a dumpsite for municipal garbage. In 1942, Hooker Chemical and Plastics Corporation drained the canal, lined it with clay, and began putting chemical wastes into it. In 1953, the company covered Love Canal with dirt and sold it to the Niagara Falls Board of Education for one dollar. Then in 1954, the Board constructed an elementary school on the site and home building began nearby. By 1978, there were approximately 800 single-family homes and 240 low-income apartments in the vicinity of the dump, as well as the Ninety-Ninth Street Elementary School.

By August 1978, a major environmental health crisis was unfolding. According to Eckardt C. Beck, regional administrator of the EPA, residents had a "disturbingly high rate of miscarriages" and Love Canal was "one of the most appalling environmental tragedies in American history."[5] The New York State Health Commissioner declared an emergency, closed the Ninety-Ninth Street School, and recommended the temporary evacuation of pregnant women and young children. At about the same time, President Jimmy Carter allocated federal funds to deal with the problem and ordered the Federal Disaster Assistance Agency to help.[6]

Love Canal was significant not only because it marked the birth of the environmental health movement, but also because it drew national attention to the health effects of toxic chemicals. When Lois Gibbs and her neighbors appeared on national TV, it made environmental issues very personal and very real. For the first time, the American public could actually see and hear the people who were being affected, and this transformed public debate. Instead of scientific and legal discussions about federal environmental policy in Washington, DC, it began to focus on real people suffering from real environmental health problems. This transformation was huge; the environment had gone from an issue debated by lobbyists and legislators inside the Beltway to one that was discussed in kitchens and living rooms across the entire country.

In the aftermath of Love Canal, Congress enacted the *Comprehensive Environmental Response, Compensation, and Liability Act* (CERCLA), otherwise known as Superfund. Superfund provided the federal government with authority to clean up abandoned toxic waste dumps. But the actual number of sites was, and still is, unknown. In 1979, the EPA estimated the number at between 32,000 and 50,000,[7] but a decade later the Office of Technology Assessment estimated there were 439,000, including federal properties, underground storage tanks, mine wastes, pesticide-contaminated land, underground injection wells, abandoned gas manufacturing facilities, wood preserving plants, and waste disposal sites,[8] but in 2012, there were about 1,300 sites listed on the Superfund National Priority List. Although much of this discrepancy may be because many sites do not pose significant risks to human health, it reveals that the exact number of toxic waste sites requiring cleanup in the United States is unknown.

Not only is there confusion about the exact number of toxic waste sites, there are huge uncertainties about how they affect human health. Despite numerous attempts to study the health problems around these sites, the federal government has had to admit its ignorance. In 1991, a National Academies of Sciences report concluded, "we find that the health of some members of the public is in danger," but "we are currently unable to answer the question of the overall impact on public health of hazardous wastes."[9] This study also reported that 4.1 million people lived within one mile of the 725 Superfund sites that were investigated. Nearly half of these were women of child-bearing age, children, or seniors, "all of whom can be considered at particular risk from toxic chemical exposures." This begs the question—if four million people lived within one mile of 725 Superfund sites, how many people live within one mile of the 439,000 sites identified by the Office of Technology Assessment?[10]

CERCLA also required the federal government to create a new agency—the Agency for Toxic Substances and Disease Registry (ATSDR)—to investigate health problems at toxic waste dumps, but it refused to act on this requirement for three long years. It was only after a 1983 lawsuit jointly filed by the Environmental Defense Fund, the Chemical Manufacturers Association, and the American Petroleum Institute that the federal government eventually complied with the law. Initially, it was unclear why industry groups would support the creation of ATSDR, but the reason soon became obvious. ATSDR came under the direct authority of Vernon Houk, who didn't believe that toxic chemicals could cause serious effects on human health. Dr. Richard Clapp, who was Director of the Massachusetts Cancer Registry, remembers attending a 1987 meeting also attended by Houk. Houk "had a dismissive attitude" according to Clapp, "maintaining that ATSDR wasn't really necessary, that there was not enough of a significant (environmental) health problem for all this attention to be paid to it."[11] Not surprisingly, most of the ATSDR reports issued under his management found no evidence of harm.[12]

Meanwhile, Lois Gibbs's victory at Love Canal transformed her from an anonymous housewife living in small-town America into a national heroine. She quickly became a source of inspiration and support for the many local environmental health and justice groups that began to spring up in communities across the country. In 1981, she established the Citizens' Clearinghouse for Hazardous Waste to provide information and resources to communities with environmental health problems. In its first five months, the Clearinghouse received over two thousand phone calls and its staff visited over sixty different communities.[13] Now called the Center for Health, Environment and Justice (CHEJ), it continues to help local communities to fight environmental health problems.

But the Citizens' Clearinghouse wasn't the only organization that supported local groups. In 1984, the National Toxics Campaign (NTC) was founded by John O'Connor, a well-known Massachusetts activist. Under his leadership, the NTC helped citizens' groups to fight toxic dumpsites and oppose garbage incinerators. In 1988, the organization established the first and only analytical laboratory in the country that was run by movement activists. During its lifetime, the NTC helped to take many polluters to court, secure passage of the 1986 reauthorization of Superfund, and create the Toxics Release Inventory. The NTC also played a key role in drawing attention to the U.S. military's environmental violations. In 1990, O'Connor coauthored the first manual on citizen activism on toxic chemicals.[14] The NTC was disbanded in 1993.

THE BEGINNINGS OF THE ENVIRONMENTAL JUSTICE MOVEMENT

Unlike the large, national mainstream environmental groups, the Citizens' Clearinghouse and the NTC were both committed to supporting poor and minority communities in their struggles for environmental health and justice. In the late 1970s, these struggles, often based in African American communities, gave rise to the new environmental justice movement. It can be difficult to separate the environmental justice and environmental health movements. Both are concerned about human health, and both are concerned about environmental injustice, but they differ in how they frame issues. The environmental justice movement foregrounds human rights and calls for justice and fairness. On the other hand, the environmental health movement puts protecting health and preventing disease front and center.

A second difference is their origins. The environmental justice movement traces its origins to the civil rights movement of the 1950s and 1960s, which in turn arose from the much earlier antislavery movement. By expanding the concept of social justice to include the right to a healthy environment, civil rights activists gave rise to a new, but related social movement. On the other hand, the environmental health movement can be traced back to the social reformers of the Industrial Revolution and their efforts to improve the living and working conditions of the urban poor. A third difference is their tactics. The environmental justice movement tends to use collective action to demand change on local issues, whereas today the environmental health movement, especially its state and national groups, tends to emphasize lobbying for new legislation. But despite these differences, these two movements have more in common than they do apart because they share a single goal—environmental health for all.

The first organized opposition to environmental injustice took place in 1979 when African American homeowners in Houston, Texas, fought to prevent construction of a garbage dump in their suburban community.[15] Residents formed a community action group which filed a class action lawsuit to block the facility from being built. The 1979 lawsuit, *Bean v. Southwestern Waste Management, Inc.*, was the first of its kind to challenge the siting of a waste facility under civil rights legislation. Although the case received relatively little media attention, it laid the groundwork for other minority communities to protest and take action on environmental injustice.

Some three years later, in 1982, local environmental justice activists were successful in attracting widespread media coverage when protests about a new PCB waste site in Warren County, North Carolina, led to five hundred arrests. Using nonviolent collective action tactics taken straight from the civil rights movement, local residents, church leaders, and activists lay down in front of trucks taking PCB-contaminated soil into this largely African American county. Although the demonstrators didn't succeed in stopping the new dump, they put the issue of environmental justice onto the nation's political agenda. As a result of the Warren County protests, the U.S. General Accounting Office (GAO) launched a study of the racial and economic status of communities close to hazardous waste sites, which showed that three of the four commercial hazardous waste sites in an eight-state region were located in African American communities even though African Americans made up only 20 percent of the population.[16] This study was the first government investigation to confirm the existence of environmental injustice in the United States.

The Warren County protests also led to a larger, national nongovernmental study. Commissioned by the United Church of Christ Commission for Racial Justice, *Toxic Waste and Race* (1987)[17] showed that race was the most important factor in the location of hazardous waste sites—more important than poverty, home ownership, or land values. But its most stunning finding was that three of every five African Americans and Hispanics lived in a community close to an unregulated site. In the wake of this disturbing information, environmental justice activists organized a mass march in 1988. Called the Louisiana Toxics March, it drew attention to the poor environmental conditions endured by African Americans living in Louisiana's "Cancer Alley," an eighty-mile corridor that's home to 150 oil refineries and petrochemical plants.

Confirming that environmental injustice is still a major problem, a 2007 follow-up report to the United Church of Christ's 1987 report stated, "Twenty years after the release of *Toxic Wastes and Race*, significant racial and socioeconomic disparities persist in the distribution of the nation's commercial hazardous waste facilities. . . . In fact, people of color are found to be more concentrated around hazardous waste facilities than previously shown."[18]

One of the leaders of the environmental justice movement is Robert Bullard. Husband of attorney Linda McKeever Bullard, who represented the Houston homeowners in 1979, he's been part of the movement since its birth. Indeed, some call him the father of environmental justice. In 1990, he published a best-selling book, *Dumping in Dixie: Race, Class, and Environmental Quality*,[19] which told the stories of environmental injustice in five African American communities. The book affected both whites and African Americans. Many white Americans were alerted to the fact that racial discrimination was still alive and well in the United States, despite the civil rights legislation of the 1960s, and many African Americans resonated with the tales of real-life communities very similar to their own. Establishing local and regional groups, they demanded fair treatment. Two of the best-known environmental justice groups established during the late 1980s and early 1990s are the West Harlem Environmental Action (WE ACT), founded in 1988, and the Deep South Center for Environmental Justice, founded in 1992. Both are still active today.

Communities protesting environmental injustice claim that their neighborhoods have been deliberately chosen for toxic dumps and other hazardous facilities because of their economic and political weaknesses. Believing that they were intentionally and unfairly targeted, these communities are justifiably angry. Two documents prove them right. One is a 1984 report prepared by Cerrell Associates, a consulting company for California's Solid Waste

Division,[20] and the other is a 1989 memo written by Eply Associates for Chem-Nuclear Systems Inc.,[21] which had been hired by the North Carolina Low-Level Radioactive Waste Management Authority to assist with the site selection and operation of a waste disposal site. In both documents, the most important siting criteria are not scientific, but demographic, and both identify communities least likely to resist.[22] They provide compelling evidence that the siting system works against politically and economically marginalized communities and that they are deliberately chosen to site facilities posing environmental health risks.

Many Native Americans communities experience environmental injustice from hazardous facilities on their lands and natural resource exploitation. Some of their lands are rich in coal, oil, gas, uranium, and gold and have been mined and drilled by major corporations for decades. Not only do the affected communities have to live with the environmental health effects, those who work for the corporations endure numerous occupational health problems; for them, it's double exposure. Tribes that have been especially affected include the Navajo/Diné (coal, uranium, oil, and gas), Hopi (coal), the Northern Cheyenne (coal and gas), the Lakota-Sioux (uranium), and the Spokane (uranium).

Perhaps the worst incident of environmental injustice against Native Americans is the 1979 disaster at Church Rock, New Mexico. On July 16, the dam at the Church Rock Uranium Mill broke, releasing eleven hundred tons of radioactive mine tailings and ninety-four million gallons of heavy metal effluent into the Puerco River, which was extensively used by the Navajo. The amount of radiation released was comparable to the Three Mile Island accident, also in 1979. But unlike Three Mile Island, this disaster was largely ignored by government agencies, the media, and scientists. Amazingly, there have been no long-term epidemiological studies of the people affected.[23] Even at the time, local residents were not informed about the disaster or the risks to their health. Would this blatant disregard for people and their health have happened in an affluent, white community? Probably not.

As evidence of environmental injustice in Native American communities mounted, it became clear that a Native presence was needed on the national stage. So just over a decade after the Church Rock disaster, in 1990, the Indigenous Environmental Network was formed. Over the years its longtime Executive Director, Tom Goldtooth, has built a strong organization comprising 250 Indigenous communities working on climate justice, energy, toxics, water, globalization and trade, and sustainable development. Since then, many other national, regional, and local Native environmental justice organizations have been established, including the National Tribal Environmental Council, the Inter-Tribal Environmental Council, and the Southwest Network for Environmental and Economic Justice. One of the most prominent Native American environmental justice groups is Honor the Earth. Founded in 1993 by Winona LaDuke, Amy Ray, and Emily Saliers, it addresses the two primary needs of the Native environmental movement: the need to break the geographic and political isolation of Native communities and the need to increase financial resources for organizing and change. LaDuke is well-known for her books—*All our Relations*[24] and *Recovering the Sacred*[25]—and for her politics: in 1996 and 2000, she ran as the vice-presidential candidate for Ralph Nader on the Green Party ticket.

In the late 1980s, African American and Native American activists realized that they'd be more effective if they worked together, so in 1991 they organized the First National People of Color Environmental Leadership Summit. Held in Washington, DC, and bringing together over 650 participants from all fifty states and several other countries, it succeeded in forging new alliances between African Americans, Native Americans, and other racial minorities. It also broadened the environmental justice movement's agenda to include worker safety, land use, transportation, housing, resource allocation, and community empowerment.[26] At the Summit's conclusion, delegates adopted a declaration articulating seventeen *Principles of Environmental Justice*.[27] These principles still provide a defining vision for environmental justice today.

By then, even the normally conservative legal profession was getting concerned about environmental injustice. In 1992, the National Law Journal released a special issue[28] which concluded that environmental cleanups were slower in communities of color and that standards were lower. It also found that EPA fines were not as strict for industries operating in communities of color. After these revelations, the federal government finally began to take the issue seriously.

The EPA's first major report on environmental injustice, *Environmental Equity: Reducing Risk for All Communities*,[29] was released later that year. Revealing that racial minorities and low-income populations had above average exposure to air pollution, hazardous wastes, contaminated fish, and agricultural pesticides, it confirmed what activists already knew—that environmental injustice was rife throughout the United States. To help reduce environmental injustice, the EPA created an Office of Environmental Equity.[30] Environmental justice activists won a further victory in 1994 when President Bill Clinton signed an executive order on *Federal Actions to Address Environmental Justice in Minority Populations and Low Income Populations*.[31] Requiring federal agencies to address the disproportionate environmental health effects of their programs on minorities and develop environmental justice strategies, it seemed like a major step forwards. But after the Clinton administration, environmental justice virtually disappeared from the federal government's agenda until a 2011 Memorandum of Understanding reaffirmed its importance.[32] Since then, the EPA seems to have regained its commitment to environmental justice.

Over the years, it's become even clearer that the environmental health and justice movements are united by a common cause—environmental health for all—even though they differ in their origins, how they frame issues, and their tactics. In recent years, these two movements are transcending their differences and working together more closely than ever before (see chapter 8).

THE ROLE OF DISASTERS IN BUILDING THE ENVIRONMENTAL HEALTH MOVEMENT

Since the events at Love Canal there have been countless environmental health disasters in the United States. Always heartbreaking and often revealing environmental injustice, they have been used by local activists to raise public awareness, organize collective action, and demand change. Sad to say, it often takes a disaster to shake people out of their everyday complacency.

In May 1979, about a year after events at Love Canal started to unfold, state investigators in Woburn, Massachusetts, discovered that two municipal wells were contaminated with the industrial solvents trichloroethylene (TCE) and perchloroethylene (PERC). Local residents, who were already concerned about high rates of childhood leukemia, cancer, and other illnesses in their community, became very alarmed and persuaded the state health department to conduct a health study. Their concerns were well-founded. The final report concluded: "Woburn's cancer death rate is higher than what should be expected."[33] Subsequent research revealed an elevated incidence of childhood leukemia and renal cancer in the town.[34] After three years of protest, in May 1982, several Woburn families filed a civil lawsuit against W. R. Grace and Beatrice Foods, the two corporations who owned the property where the contamination seemed to have originated. The jury found Beatrice not guilty, however it came to the opposite decision about Grace's culpability. Subsequently, the judge threw out the jury's decision against Grace on the grounds that it was confusing and contradictory. But after the verdict, the EPA said that both corporations were responsible, and the site is now being cleaned up under the Superfund program. Later, a book called *A Civil Action*[35] was written about this disaster, and in 1998 it was turned into a movie starring John Travolta.

Like Love Canal and Woburn, Times Beach, Missouri, was an undistinguished small town until its legacy of pollution became national news. In November 1982, almost a decade after the town hired Russell Bliss to spray waste oil on local roads to reduce dust levels, it became apparent that the oil contained dioxins. After an announcement from the EPA, the community went into shock. According to Marilyn Leistner, the Mayor of Times Beach:

> Chaos broke loose. The residents immediately recalled that the roads had turned purple after being sprayed. The spraying had resulted in an awful odor.

> Birds had died and newborn animals succumbed shortly after their birth. One man remembered a dog found in one of the contaminated ditches. They thought the dog rabid and prevailed upon a policemen to shoot it. Another man told how he had called the St. Louis Health Department to tell them about the dead birds he kept finding. The department recommended that he freeze the dead birds and said they would be out to pick them up. No one ever came.[36]

To make matters worse, residents were told they would have to wait nine months for government test results. Unwilling to endure this delay, the community decided to take matters into its own hands and to pay for private testing immediately. On December 13, preliminary results showed that the whole town was contaminated with dioxins and ten days later Times Beach was evacuated. On February 23, 1983, the EPA announced a voluntary buy-out of the entire town for $32 million. Today, Times Beach is a state park. The environmental health disaster at Times Beach remains the largest public exposure to dioxins in the United States. Partly as a result of events in this small town, the *Comprehensive Environmental Response, Compensation, and Liability Act* (Superfund) was strengthened and replaced by the *Superfund Amendments and Reauthorization Act* (1986).

Since then, there have been numerous other environmental health disasters in the United States, including drinking water contamination in Hinckley, California; asbestos contamination in Libby, Montana; PCB poisoning in Anniston, Alabama; a coal sludge spill in 2000 in West Virginia and Kentucky; Hurricane Katrina in 2005; lead contamination in Picher, Oklahoma; and the Tennessee coal ash spill of 2008—to name but a few. Although they have harmed human health and well-being, something positive has come out of them, namely, strong and effective community-based activism on environmental health. In fact, it's always local residents who demand action because they are the most affected. Most do this work as unpaid volunteers and learn on-the-job how to be community organizers, fundraisers, researchers, and media spokespeople. Many are victims themselves and are suffering the loss of their health, their homes, and their quality of life. But despite this, and perhaps because of it, local environmental health activists are tough, resilient, and determined to succeed. They are an example of philosopher Frederick Nietzsche's statement that "what doesn't kill me makes me stronger." Indeed, local campaigners have always played a leading role in the struggle for environmental health in the United States. Without them, there wouldn't be a social movement.

STRUGGLES FOR REGIONAL ENVIRONMENTAL HEALTH IN THE GREAT LAKES

But it's not only local disasters and local activists that have played an important role in the U.S. environmental health movement; it's also regional problems and regional activism, such as that in the Great Lakes. The Great Lakes is one of the most heavily populated and industrialized regions in North America. About 33.5 million people call it home, including 25 million Americans and 8.5 million Canadians. On top of that, the region was one of the first to industrialize in North America. Attracted by cheap hydroelectric power and an abundance of water and natural resources, many industries set up shop in the Great Lakes, including steel, paper, chemicals, and automobiles. Not surprisingly, this combination of population and industry has led to a host of environmental health problems.

The first recognition of serious problems came in the late 1960s, when Lake Erie was pronounced dead (see chapter 3). This spurred the development of the first *Great Lakes Water Quality Agreement* between Canada and the United States in 1972. This *Agreement* committed the two federal governments to restore and enhance water quality in the region. It also established water quality objectives and agreement to codesign and implement monitoring programs. Intended to control phosphorus and other nutrients that had caused the death of Lake Erie, it marked an important precedent for binational cooperation, even though it did not specifically address human health.

It soon became clear that pollution wasn't limited to Lake Erie or to nutrients and that it was a problem throughout the entire region. In response, U.S. environmental groups swung into action and demanded stronger measures to protect the Great Lakes. By the mid-1970s, several national groups had established campaigns, including the National Audubon Society, the Izaak Walton League, the Sierra Club, and the League of Women Voters. Regional groups played an important role too, including the Lake Michigan Federation and the Michigan United Conservation Clubs. In Canada, Pollution Probe, the Canadian Institute for Environmental Law and Policy, the Canadian Environmental Law Association, and the Société pour Vaincre la Pollution all developed Great Lakes campaigns over time.

During the 1970s and 1980s, studies showed that all of the Great Lakes were heavily contaminated with persistent toxic chemicals. Similar to the antinuclear movement of the 1950s and 1960s, it was scientists who raised the alarm and provided environmental groups with the ammunition they needed to make their case. Key among them were Canadians Jack Vallentyne, Henry Regier, and Doug Hallett and Americans Al Beeton, Wayland Swain, and Bill Sonzogni. By 1978, they had identified four hundred contaminants in the air, water, land, and living organisms of the region.[37] Then, in the 1980s, Jo and Sandra Jacobson at Wayne State University in Detroit and their coresearchers began publishing a series of scientific papers on the effects of maternal consumption of PCB-contaminated fish on the development of infants and young children.[38] These papers were among the first to document the health effects of toxic chemicals on ordinary people, and the first to show that environmental contaminants could affect the children of exposed women.

In 1978, the rising level of scientific and public concern prompted the United States and Canada to revise the 1972 *Great Lakes Water Quality Agreement.* The 1978 *Agreement* went further than the earlier one and provided a framework for binational efforts to rid the Great Lakes of toxic chemicals forever. It was a visionary document. Requiring the virtual elimination of persistent toxic substances, the *Great Lakes Water Quality Agreement* (1978) stated: "The discharge of toxic substances in toxic amounts (shall) be prohibited and the discharge of any and all persistent toxic substances (shall) be virtually eliminated."[39] Ever since then, "virtual elimination" has been one of the demands of the environmental health movement.

Not satisfied with the lack of progress, in 1982 groups in Canada and the United States formed a new binational environmental health coalition—Great Lakes United.[40] This organization is important not only because it was binational, but also because it was one of the first regional coalitions on environmental health. Today, Great Lakes United is still a well-respected champion for protecting the Great Lakes.

With the encouragement of Great Lakes United and its coalition partners, the Canadian and U.S. governments further strengthened the *Agreement* in 1987. In an unusually progressive move, the two federal governments invited environmental groups to send representatives to sit in on the final negotiations and to participate in some of the discussions. I was fortunate enough to be chosen as a representative of Great Lakes United.

With the 1987 amendments to the *Great Lakes Water Quality Agreement*, environmental health became a critical issue in the Great Lakes basin. A 1992 report published by the International Joint Commission (IJC), the binational governmental organization responsible for monitoring progress towards the *Agreement*'s goals, stated:

> It is clear to us that persistent toxic substances have caused widespread injury to the environment and to human health. As a society, we can no longer afford to tolerate their presence in our environment and in our bodies. Their use and presence in the Great Lakes environment are also inherently inconsistent with the Agreement's purpose and specific provisions.[41]

Just read those words again and remember that they were written by an agency of the U.S. and Canadian governments—not radical activists. The strength of this language is truly astonishing. Not surprisingly, the 1992 IJC report received considerable media attention. Rachel's Hazardous Waste News enthusiastically reported:

> An important breakthrough in control of toxics occurred during April. The International Joint Commission (IJC) . . . made far-reaching official recommendations which, for the first time, embody a truly modern approach to the identification and control of toxic chemicals. It appears to be a real first step toward a sustainable world.[42]

The year 1992 marked the high point for environmental health concerns in the Great Lakes. In the years that followed, the two federal governments gradually lost interest and increasingly ignored the IJC's recommendations. Increasingly preoccupied with trade liberalization, they signed the North American Free Trade Agreement (NAFTA) with Mexico in 1993, effectively sidelining the *Great Lakes Water Quality Agreement* and the IJC. Then in September 2012, the U.S. and Canadian governments initialed another amendment to the *Great Lakes Water Quality Agreement*. This one is a pale imitation of the earlier ones. Although it covers some issues not addressed previously, like climate disruption, it is long on promises and short on specific goals and actions. Even worse, it pays little attention to human health.

Although the promise of the *Great Lakes Water Quality Agreement* has not been fulfilled, the work done by environmental health activists in the region has rippled out across the country and internationally. They helped to raise awareness about the health problems caused by toxic chemicals, and their efforts to achieve "virtual elimination" inspired the entire movement, as well as governments around the world. In fact today, the concept of virtual elimination can be found in international environmental agreements, national legislation, and state and local regulations.

WINNING THE BATTLE AGAINST WASTE INCINERATION

One of the urban environmental issues that worried environmental health activists in the Great Lakes and elsewhere was garbage incineration. By the late 1970s and 1980s, municipal waste sites across the United States were filling up and many cities and towns were refusing to accept the construction of new ones or the expansion of existing ones. Using the slogan "Not In My Back Yard" (NIMBY), local activists successfully opposed many proposals. At the same time, garbage disposal was getting more expensive because of the requirements of the *Resource Conservation and Recovery Act* (1976). The combination of these two factors forced many municipalities to look for alternatives to land-based garbage disposal.

Waste incineration quickly became an attractive option, prompting studies into the feasibility of "energy-from-waste" plants, similar to those already operating in Europe. The idea of waste incineration was strongly endorsed by the four big nuclear power plant manufacturers—Combustion Engineering, Babcock & Wilcox, General Electric, and Westinghouse—who were looking for a new line of work after the 1979 Three Mile Island accident ended construction of nuclear power plants in the United States.

The Department of Energy (DOE) also became a supporter, believing that waste incineration could help the United States meet its growing energy needs and secure energy independence. According to a 1980 report, the Department's goal was to build between 200 and 250 new incinerators by 1992 that would burn 75 percent of the nation's garbage. Initially, it seemed that this goal would become a reality, and in 1985, forty-two new incinerators were ordered, but by 1987 this number had shrunk to twenty-five. In fact in 1987, more capacity was canceled (35,656 tons per day) than was ordered (20,585 tons per day). The incineration industry failed.[43] Why?

Because community groups successfully opposed it. Concerned about hazardous air pollutants, the disposal of toxic ash, process inefficiencies, and high costs, they protested the construction of new waste incinerators. They also fought to close existing ones, most of which lacked adequate pollution controls and discharged large amounts of toxic chemicals into the air. But local campaigners didn't just oppose incineration, they also developed alternative waste management strategies—most notably curbside recycling. In the late 1980s and 1990s, programs were launched in thousands of municipalities across the United States, resulting in dramatically higher recycling rates. Between 1980 and 2007, the national recycling rate increased from less than 10 percent to over 33 percent. At the same time, the proportion of waste disposed of at garbage dumps decreased from 89 percent to 54 percent.[44] The work of local activists to develop alternatives to waste incineration is hugely important because it showed that they could develop constructive solutions to environmental health problems, as well as be critical of the *status quo*.

As community activists successfully closed many old, polluting municipal waste incinerators, it became clear that another type of incinerator was a major contributor to toxic air pollution. In the 1990s, studies showed that the wastes burned by hospital incinerators were the leading source of dioxins to the atmosphere. The local groups that had worked on municipal waste incineration now turned their attention to this problem and achieved similar success.

They were supported by a new national coalition called Health Care Without Harm (HCWH). Initially comprising twenty-eight organizations, it was formed in 1996 by Gary Cohen—former Executive Director of the National Toxics Campaign—Charlotte Brody, Jackie Hunt Christensen, and others. HCWH started out as a coalition to advocate for the closure of hospital incinerators, and it was highly successful, helping to close thousands across the United States and other countries. Today, HCWH is a broadly based international coalition of almost five hundred organizations in more than fifty countries working to reduce the use of toxics in health care products, and on climate disruption, green building, energy, healthy food systems, and other issues (see chapter 9).

OPPOSITION TO PESTICIDES: AN ONGOING STRUGGLE

Pesticides were another important issue for the young U.S. environmental health movement, and they've remained on the agenda ever since. Although 80 percent of pesticides used in the United States are for agriculture, they're very widely used in the urban environment. Acre for acre and pound for pound, pesticide use can be greater in cities than in the countryside. According to the EPA, over 170 million pounds were used in government, industrial, commercial, and home and garden applications in 2007.[45] In particular, golf courses, public parks, cemeteries, and gardens are often saturated with frequent applications, especially during the summer months when people are outside. But in addition to their use in urban settings, there are many other concerns about pesticides, including farm worker exposure and consumer exposure through food. Put it all together, and every man, woman, and child in America is exposed to pesticides every day of their lives.

Early concerns about pesticides focused on their effects on wildlife. Heeding the warning in *Silent Spring*, the environmental movement began to advocate for stronger controls on DDT in the late 1960s. More concerned about its effects on bald eagles, ospreys, and other raptors than its effects on human health, they lobbied for a ban, and eventually, in 1972 they were successful. At about the same time, a few agricultural scientists were becoming alarmed. Increasing pesticide use was causing the spread of pesticide-resistant insects, air and water pollution, and the death of "nontarget" wildlife species. Robert van den Bosch, a professor of entomology at the University of California, expressed his concerns saying, "Pesticides rank with the most dangerous and ecologically disruptive materials known to science, yet under the prevailing system these biocides are scattered like dust in the environment."[46] A father of integrated pest management (IPM), which promotes safer means for controlling pests, van den Bosch argued that pesticides should only be used when absolutely necessary.

By the late 1970s, concerns about the effects of pesticides on human health became more prominent. In 1977, the Northwest Coalition for Alternatives to Pesticides was formed in Oregon, followed in 1981 by the National Coalition Against the Misuse of Pesticides, since renamed Beyond Pesticides. These organizations opposed pesticides and advocated IPM. Then, in

1982, U.S. pesticide activists met with their colleagues from other countries to discuss the increasing global trade and its consequences for human health. Concerned about what they had read in a book called *Circle of Poison: Pesticides and People in a Hungry World*,[47] they established the international Pesticide Action Network (PAN) to challenge the global proliferation of these toxic substances.

Initially, PAN focused on lobbying the UN Food and Agriculture Organization (FAO) to adopt Prior Informed Consent for pesticides—the policy that requires countries exporting hazardous substances to notify importing countries and provide them with information. But in 1985, PAN broadened its approach and launched the Dirty Dozen campaign, which advocated a ban on twelve of the most toxic organochlorine pesticides: aldicarb; toxaphene; chlordane and heptachlor; chlordimeform; chlorobenzilate; dibromo-3-chloropropane (DBCP); DDT; aldrin/dieldrin and endrin; ethylene dibromide (EDB), hexachlorocyclohexane (HCH) and lindane; paraquat, parathion and methyl parathion; pentachlorophenol; and 2,4,5-T. Eventually, its work led to an international environmental agreement—the 2001 Stockholm Convention on Persistent Organic Pollutants (POPs)—which banned or restricted the Dirty Dozen.

In 1984, pesticides became an important labor issue in the United States when Cesar Chavez (1927–1993), founder and leader of the United Farm Workers (UFW), launched the Grapes of Wrath campaign. Arguing that exposure to pesticides caused illness among farm workers, especially migrant Latinos, the campaign urged a boycott of California table grapes until growers stopped using five of the worst ones. The UFW ended the Grapes of Wrath campaign in 2000, after three of the targeted pesticides were banned. The Grapes of Wrath campaign is significant because it was probably the first national campaign to directly link environmental and occupational health. Drawing on the name of John Steinbeck's well-known novel about migrant agricultural workers in the 1930s Dust Bowl, the Grapes of Wrath campaign also represented one of the first market campaigns on environmental health (see chapter 9).

Today, farm workers are still directly exposed through spraying and handling treated fruit, vegetables, and other produce, and their families are still affected by pesticide drift and by workers bringing home residues on their clothes and shoes. Organizations like Farmworker Justice and the Farm Worker Pesticide Project (FWPP), which understand the link between environmental and occupational health and justice, work with affected populations and advocate for reducing pesticide use.

But it's not only farm workers and their families that face risks from agricultural pesticides—it's also consumers and their children. This became obvious after the release of a report called *Intolerable Risk: Pesticides in our Children's Food*[48] by the Natural Resource Defense Council (NRDC). Published in February 1989, it targeted Alar, a carcinogenic chemical used on apples and other fruit. The report used government data to show that young children faced "an intolerable risk" from a wide variety of chemicals in their food, including Alar. According to the NRDC, "the average pre-schooler's exposure was estimated to result in a cancer risk 240 times greater than the cancer risk considered acceptable by the EPA following a full lifetime of exposure." The report was featured on CBS's *60 Minutes*, and soon after the EPA announced a ban on Alar.

A fierce backlash followed publication of the NRDC report and the *60 Minutes* feature. The American Council on Science and Health (ACSH), which had received $25,000 from Alar's manufacturer,[49] disputed that Alar and UDMH, its breakdown product, were harmful. Claiming that the "Alar scare" was completely unwarranted, the ACSH was quite successful at dissipating public concern. Even today the phrase "Alar scare" is used to describe news stories supposedly based on scare tactics, emotion, and propaganda. But the "Alar scare" was no scare. Alar is classified as a probable human carcinogen by the EPA, and its breakdown product UDMH is classified as a possible carcinogen by the International Agency for Research on Cancer.

The NRDC report drew attention to children's environmental health and turned it into a hugely important issue for the new environmental health movement (see chapter 6). In 1993, the National Research Council (NRC) published its own study on *Pesticides in the Diets of Infants and Small Children.*[50] Agreeing with the findings of the NRDC report, it concluded that regulations on pesticide residues in food did not adequately protect infants and children. The NRC report demonstrated that infants and children face much greater risks because their diets are significantly different from adults' and because they are more vulnerable to the effects of toxic chemicals. In doing so, it legitimized and extended the environmental health movement's concerns about children's health.

With this ammunition in hand, activists and parents lobbied Congress to protect children's environmental health, and in 1996, legislators took action and passed the 1996 *Food Quality Protection Act.* Although this *Act* provided an extra safety margin in the standards for pesticide residues in food to protect children's health, its wording was unclear. It was challenged in court and has never been completely implemented. The *Food Quality Protection Act* also repealed the application of the Delaney Clause to carcinogenic pesticide residues in food. The victory that the environmental health movement had hoped for turned out to be rather hollow.

But environmental health activists didn't give up. In 1999, a decade after the NRDC's original study, the Environmental Working Group (EWG) published a report called *How 'Bout Them Apples? Pesticides in Children's Food Ten Years after Alar.*[51] EWG used the federal government's own food consumption data and found that the gap between the health risks to children and adults was getting larger, not smaller. Since then, EWG has published several consumer guides on pesticide contamination in food and advocated for stronger pesticide legislation. The work of EWG and other environmental health groups has done an enormous amount to raise public awareness about pesticides in food and change consumer purchasing choices.

Pesticides have always been one of the hallmark issues of the U.S. environmental health movement and a bridge with the environmental justice movement. Pesticides are important because everyone is exposed. Whether it's migrant farm workers spraying pesticides and handling treated produce, their families exposed to drift, or children eating contaminated food, or public exposure in the urban environment, we are all at risk. For years, the U.S. environmental health movement has opposed pesticide use, and although the struggle continues, much has been achieved to control this type of toxic substance and educate the public.

SECURING THE RIGHT TO KNOW

As Americans became more aware about pesticides and other hazardous substances in the environment, they began to demand the "right to know" about the toxic chemicals in their communities. They wanted to know which ones were present and how they could affect health. Receiving no satisfactory answers from government or industry officials, they started to lobby for the right to know this information. Concerned citizens and environmental health campaigners joined forces with workers, who were already advocating for the right to know about chemicals in their workplaces. This alliance was strongest in the heavily unionized and progressive cities of the Northeast and Midwest and in California. However, in 1981 an East Coast city—Philadelphia—became the first to enact right-to-know legislation. Cincinnati was a close second. In California, the Silicon Valley Toxics Coalition led a successful campaign targeting cities in the Santa Clara valley, and in New Jersey, hundreds of labor unions, firefighters, environmental and community organizations, churches, and women's groups became actively involved in a campaign for a state right-to-know law. In an unlikely partnership, antinuclear activists and nuclear power plant workers joined forces, as did the Fraternal Order of Police and the American Civil Liberties Union.[52] In 1983, they were successful and New Jersey became the first state to pass right-to-know legislation. Massachusetts followed in 1985.

In 1984, an environmental health disaster halfway around the world in Bhopal, India, added to the mounting pressure for federal right-to-know legislation. In the early hours of December 3 a pesticide manufacturing plant, owned by a subsidiary of U.S. chemical giant Union Carbide, released forty-two tons of methyl isocyanate (MIC) into the atmosphere. At least 520,000 people were exposed to the toxic gas. Within two weeks an estimated eight thousand people were dead, and since then, an additional eight thousand have died.[53] A total of 170,000 people were treated at hospitals and temporary dispensaries, and many more suffered illnesses. The tragedy in Bhopal remains the world's worst chemical accident. Even over twenty-five years later, there are many lingering health problems.

This disaster and others in the United States, combined with the new state and local right-to-know legislation and lobbying from environmental health groups, led to passage of the *Emergency Planning and Community Right-to-Know Act* (EPCRA) in 1986. Contained in Title III of the *Superfund Amendments and Reauthorization Act*, the dual purposes of EPCRA are emergency planning and community right to know. The *Act* requires businesses to report their emissions of specific chemicals every year, and then this information is made publicly available through the Toxics Release Inventory (TRI). Today, TRI covers some 650 potentially toxic chemicals used in almost twenty-one thousand facilities across the United States. But, as the EPA acknowledged in its report on the 2011 data, the database provides only a snapshot of the pollution produced by American industry: "Users of TRI data should be aware that . . . it does not cover all toxic chemicals or all sectors of the U.S. economy. . . . Furthermore, the quantities of chemicals reported to TRI are self-reported by facilities."[54]

When introducing the concept of TRI to the Senate, Senator Robert Stafford (R) claimed it would "reveal geographic and industrial patterns of environmental release, which health officials can correlate with records of disease incidence to seek out possible relationships."[55] Although it's true that TRI provides useful information, it has not lived up to this promise. One reason is that it takes at least two years for the EPA to release TRI information, so the public must wait to find out what they've been exposed to. A second and more important reason is that information on toxic releases and information on geographic patterns of disease have not yet been adequately correlated, although this was one of the original demands of right-to-know activists. It's true that people can map the location of facilities that report their emissions to TRI with some information on health risks and some actual health data, and that the EPA's Risk Screening Environmental Indicators Project and the National Environmental Public Health Tracking Program have tried to link environmental and health information. But the reality is that neither health officials nor environmental officials have systematically connected the size and types of industrial releases with actual patterns of disease. As Sandra Steingraber says in her book *Living Downstream*, "we track pizza deliveries and overnight packages better than we track toxic chemicals and cancer diagnoses."[56]

Today, right-to-know activists continue to advocate for more information. Many of their concerns have not changed since the 1980s, including the need for greater access to data on chemical releases and transfers, accidents and spills, hazardous wastes and permit violations. But the world has changed radically since then. Now that electronic technologies have made information collection, analysis, and dissemination much easier, government agencies and corporations have fewer reasons to avoid public disclosure, however, some information about chemical and nuclear facilities is now more difficult to access because of fears about terrorism.

Securing the right to know was an important victory for the U.S. environmental health movement. Even though it's still difficult to correlate pollution with actual disease rates, information about chemical releases has provided local and state groups with the ammunition they need to organize collective action and demand change.

TOXICS USE REDUCTION AND POLLUTION PREVENTION: LIMITED SUCCESS

By 1987, it was obvious that the environmental legislation of the 1970s was not sufficient to protect environmental health, even when supplemented with the right to know. Technology-based discharge limits were only a partial solution, and changes at the front end of the industrial production system were urgently needed. This new idea was called "toxics use reduction." Developed by the National Toxics Campaign and other groups, it advocated the redesign of manufacturing processes to encourage the use of safer feedstock materials, clean production methods, and the manufacture of nontoxic products. Toxics use reduction revolutionized thinking about the management of toxic substances and promoted strategies to control them at all stages of industrial production—"from cradle to grave" as it was called.

Encouraged by local and state groups, Massachusetts became the first state to pass toxics use reduction legislation. The Massachusetts *Toxics Use Reduction Act* (1989) set a goal of reducing the generation of toxic wastes by 50 percent and declared that toxics use reduction was the preferred means for achieving compliance with environmental and occupational health legislation. It also established the Toxics Use Reduction Institute (TURI) on the Lowell campus of the University of Massachusetts. Providing research, education, and training, TURI is an invaluable resource for industries in Massachusetts and for the entire environmental health movement. In 2009 it lost much of its state funding, despite generating significant revenues for the state's coffers.

One might assume that the labor movement would have been a strong supporter of toxics use reduction. After all, reducing workers' exposure to toxics could significantly improve occupational health. But although the Massachusetts Committee on Occupational Safety and Health (MassCOSH) encouraged unions to get involved in the subject,[57] organized labor failed to wholeheartedly support toxics use reduction at the national level. This situation highlights the fact that there was very little sustained collaboration between the U.S. environmental health movement and organized labor at the national level in the 1980s and 1990s. Cooperating on individual issues, such as the right to know, was regarded as beneficial, but ongoing, strategic alliances weren't seen as a priority by either side. This was largely a consequence of the historically poor relationships between the environmental movement and organized labor (see chapter 3).

For this and other reasons, efforts on toxics use reduction at the federal level had limited success. For instance, the federal government decided to use the term "pollution prevention" instead of "toxics use reduction." A softer and less politically charged term, it lacked the force of the phrase used by environmental health activists. Moreover, the federal government only recommended voluntary measures for pollution prevention in its *Pollution Prevention Act* (1990). While toxics use reduction was viewed as a regulatory tool, pollution prevention merely encouraged corporations to prevent pollution. Focusing on optional planning measures, the *Pollution Prevention Act* was in stark contrast to the regulatory controls imposed by the environmental legislation of the 1970s.

Looking back, it's clear that the potential of toxics use reduction has not been fully realized. Despite the efforts of many U.S. environmental health groups and some support from labor, the lack of strong federal policy has made it impossible to achieve across-the-board success. However, toxics use reduction has been important in the struggle for environmental health because it represented a shift in thinking about how to reduce risks. It also provided the foundation for later work on green chemistry and safer materials (see chapter 9).

THE LEAD SAGA

The U.S. environmental health movement's work on specific toxic substances has been more successful than its efforts on toxics use reduction. Lead is a good example. Until the 1970s and 1980s, when environmental health concerns led to the imposition of regulatory controls, exposure to this toxic heavy metal seriously damaged the health of millions of American children. A known neurotoxicant, even very low doses can impair a child's intellectual development, behavior, and learning. Lead can also cause anemia, appetite loss, abdominal pain, constipation, loss of hearing, fatigue, sleeplessness, irritability, and headaches.

Lead has been widely used for centuries despite its known toxicity. The Romans used it for making water pipes, cooking equipment, serving dishes, coins, cosmetics, and paint, even though they knew it was "injurious to the human system."[58] Lead was also widely utilized during the Middle Ages and the Industrial Revolution. By the twentieth century, it was used in paints, gasoline, water pipes, solder, lead-acid batteries, roof construction, ammunition, radiation shielding, ceramic glazes, and pesticides. Because it has excellent vibration-dampening characteristics, lead was even built into the foundations of the Pan Am Building in New York constructed next to Grand Central Station in the early 1960s.

In the early twentieth century, the hazards of lead paint were becoming apparent. In 1913, Alice Hamilton warned about its dangers saying, "the danger from the use of lead paints comes from paint dust in the air and from paint smeared on the hands which may be carried into the mouth with food or tobacco."[59] In the 1920s, lead paint was banned from indoor use by several European countries, including Austria, Belgium, Czechoslovakia, France, Greece, Poland, Spain, and Yugoslavia, but in the United States its health effects were ignored or downplayed. In fact, the National Lead Company and Dutch Boy, a prominent paint manufacturer, launched an aggressive advertising campaign to convince the public that lead was safe. As described in Markowitz and Rosner's book *Deceit and Denial*,[60] this campaign was deliberately targeted at children and their parents. One promotional jiggle proclaimed:

> The girl and boy felt very blue
> Their toys were old and shabby too,
> They couldn't play in such a place,
> The room was really a disgrace.
> But all at once they chanced to spy
> The Dutch Boy painter passing by.
> "Oh Mother!" each one cried with joy,
> "Please let us play with that nice boy!" . . .
> "This famous Dutch Boy lead of mine
> Can make this playroom fairly shine
> Let's start our painting right away
> You'll find that work is only play."[61]

The industry's campaign was successful, and lead was introduced into even more consumer products. It wasn't until 1977, more than fifty years after it had been banned in many European countries, that the Consumer Product Safety Commission announced a final ban on lead-containing paint. Even today, many children, especially those living in older homes, continue to be exposed to lead from old, peeling paint.

But lead paint is only one source of exposure; drinking water is another. Lead water pipes were common throughout the United States until the early twentieth century, and until relatively recently lead was used to solder the service connections that link homes to public water supplies. However, the warning signs were present much earlier. According to Werner Troesken's book *The Great Lead Water Pipe Disaster*,[62] the first warning came in an 1868 editorial in the *New York Herald* written shortly after several people became ill and died from lead-contaminated drinking water. It wasn't until 1991, after pressure from environmental health groups, that the federal government regulated levels of lead in drinking water in homes. Despite this, lead-contaminated drinking water continues to be a problem in homes with lead service connections and in some schools.

Although lead in paint and drinking water were important sources of exposure, by far the largest single source for most of the twentieth century was leaded gasoline. In the early 1920s, U.S. automotive companies were searching for a fuel additive to improve performance and reduce engine knock. The answer came in 1922, when two General Motors (GM) engineers—Thomas Midgeley and Thomas Boyd—announced that tetraethyl lead (TEL) could do the job. The following year, the first gasoline containing TEL was sold to the public. Unfortunately, TEL created more problems than it solved. The first warning was a strange illness that forced Thomas Midgeley to take time off work in the winter of 1923. He had been experimenting with different methods of manufacturing TEL and did not realize how dangerous it was. Then in 1924, fifteen workers became ill and died at refineries using TEL in New Jersey and Ohio. This led to a temporary ban on production and the appointment of an expert panel to assess the health risks. With the exception of Alice Hamilton, the panel was stacked with industry scientists. So it was no surprise when it concluded there were "no good grounds for prohibiting the use of ethyl gasoline . . . as a motor fuel."[63]

Over the next fifty years, the regulatory tide slowly turned against lead. In January 1971, the EPA's first administrator, William D. Ruckelshaus, declared, "an extensive body of information exists which indicates that the addition of alkyl lead to gasoline . . . results in lead particles that pose a threat to public health."[64] Then in 1973, the EPA released a report stating that lead from automobile exhaust was a direct threat to health[65] and less than a month later, the agency issued regulations requiring a gradual reduction in the lead content of gasoline over a period of five years. Ethyl Corporation, the principal manufacturer of TEL, tried to sue, but the regulations were upheld in court. In 1986, the EPA further reduced the amount of lead permitted in gasoline, and in 1996 the sale of leaded gasoline was banned for most uses.

The federal government banned the sale of leaded gasoline only after lobbying from environmental health groups and scientists. The Environmental Defense Fund (EDF) and its senior scientist at the time, Ellen Silbergeld, played a leading role. Relying on the growing body of scientific evidence showing that lead affected children's health, Silbergeld and others made a strong case for regulatory reform. Research conducted by Dr. Herb Needleman and his colleagues at the University of Pittsburgh was especially significant. One of their first studies was published in 1979. Following in the footsteps of Barry Commoner's Baby Tooth Survey and studying baby's teeth, they correlated lead levels with the behavior, IQ, and school performance of Boston-area schoolchildren.[66] Then in 1984 Needleman and David Ballinger estimated that almost 680,000 children under the age of six years old were "lead intoxicated,"[67] and in 1996 Needleman and his colleagues showed that lead is associated with aggressive behavior, delinquency, and attention deficit disorder.[68] Although their findings were attacked by industry-supported scientists, and Herb Needleman almost lost his university position, they have been confirmed by many others including Bernie Weiss and Bruce Lanphear.

There's no doubt that the elimination of lead from gasoline has significantly reduced children's exposures. Results from the National Health and Nutrition Examination Surveys (NHANES) show that between 1976–1980 and 1999–2002, the proportion of children aged one to five years old with blood lead levels higher than the CDC's action level dropped from 77.8 percent to 1.6 percent.[69] This is all the more notable because in 1991 the Centers for Disease Control (CDC) reduced its action level for lead in blood from twenty-five micrograms per deciliter to ten micrograms per deciliter.[70] However, recent studies show that there is no safe level of exposure to lead. In response, in 2012 the CDC changed its position and now recommends that a "reference value," based on the 97.5th percentile of the blood lead level distribution in children one to five years old (currently five micrograms per deciliter), be used to identify children with elevated blood lead levels.[71] There are approximately 450,000 U.S. children with levels above this cutoff value. It's clear that a huge number of children still face significant health risks, especially African Americans and other minority populations, those who live in older homes, and those who are members of low-income families.

Although the campaign against lead is a success story for the U.S. environmental health movement, it's not an unqualified one. Even though there was ample evidence about the health effects of lead by the early twentieth century, it took decades to persuade U.S. government agencies to phase out its most hazardous uses, and during this time, many millions of children were unnecessarily exposed and suffered serious health consequences. To add to this, children are still exposed through old lead paint and drinking water, and through toys, especially those imported from some countries in Asia.

NEWER ISSUES: ENDOCRINE DISRUPTORS AND EPIGENETICS

In the early 1990s, environmental health scientists began to study a new problem caused by toxic chemicals—disruption of the endocrine hormone system. Research had already shown that toxic chemicals could cause a variety of health effects, including various forms of cancer, birth defects, genetic mutations, and effects on children's behavior and IQ. But now a different and more insidious problem was becoming obvious. The Jacobson's research in the Great Lakes on the growth and developmental effects of maternal consumption of contaminated fish on infants and young children captured the attention of many scientists, including zoologist Theo Colborn. A researcher at the World Wildlife Fund in Washington, DC, she began to collect and summarize studies on how toxics affected the growth and development of wildlife. She soon noticed that health problems in wildlife were very similar to those being reported by the Jacobsons in human children in the Great Lakes. Could there be a connection between them? Could the health problems in wildlife and children be caused by the same thing?

Recalling that the drug diethylstilbestrol (DES), given to prevent miscarriages, had turned out to cause cancer and other health effects in the children of mothers who took the drug, Colborn speculated that the health effects she was witnessing in wildlife and humans were probably caused by prenatal exposure to toxic chemicals. She went on to propose that toxic chemicals exerted these effects by disrupting the metabolism of endocrine hormones, such as estrogen. In other words, not only could toxic chemicals cause health effects in adult wildlife and humans, they could interfere with hormone metabolism and harm offspring before they were even born.

Today, we know there are many different types of endocrine disrupting substances, including hormones, plant constituents, pesticides, pollutants, and compounds used to make plastics. Some are persistent organic pollutants (POPs) that do not break down in the environment and can be transported thousands of miles by air currents and migrating wildlife. We also know that endocrine disrupting substances can cause many different health effects, including reduced fertility; male and female reproductive tract abnormalities: skewed male/female sex ratios; loss of fetus; menstrual problems; changes in hormone levels; early puberty; brain, behavior, and learning problems; impaired immune functions; and various forms of cancer.

As research on these substances has progressed, it has also become clear that the type and severity of their effects varies depending on precisely when exposure happens. Prenatal exposure is especially hazardous because developing organisms are extremely sensitive. The finding that the precise timing of exposure influences the effects was stunning because it overturned the toxicological dictum that "the dose makes the poison" and the belief that there is always a clear relationship between dose and effect—a biological gradient. The phenomenon of endocrine disruption has shown that when an exposure takes place can be more important than its magnitude.

Theo Colborn and her colleagues published the first scientific paper on endocrine disruption in 1993,[72] and in 1996 she coauthored a book on the subject with Dianne Dumanoski and John Peterson Myers called *Our Stolen Future*,[73] (Myers, as Executive Director of the W. Alton Jones Foundation, had funded some of Colborn's early work). There are now thousands of articles and books, as well as government policies and research programs, on endocrine disruptors, and Colborn's hypothesis has become widely accepted. She has won many national and international awards for her work and is widely recognized as a pioneer in environmental health.

In recent years, the science of endocrine disruptors has been linked to the developing discipline of epigenetics. Epigenetics is the study of changes in gene expression without genetic mutation. What's particularly alarming is that these changes can often be passed from one generation to the next, like the developmental effects of endocrine disruptors. Conrad Waddington (1905–1975) was one of the first scientists to postulate that genes could be affected by the environment. Before then, scientists had assumed that genetic mutation was the only process that could affect gene expression, but Waddington suggested that the environment itself could play a role, and although he didn't suggest a mechanism, he was right. Today, there are several known epigenetic mechanisms including DNA methylation, histone modification, and microRNA expression.[74] Moreover, scientists have identified several classes of toxic chemicals that exhibit epigenetic activity, including heavy metals (cadmium, arsenic, nickel, chromium, methylmercury), peroxisome proliferators (trichloroethylene, dichloroacetic acid, trichloroacetic acid), air pollutants (particulate matter, black carbon, benzene), and endocrine-disrupters (diethylstilbestrol, bisphenol A, persistent organic pollutants, dioxins). This alarming discovery suggests that exposure to toxic chemicals may pose even greater health risks than previously believed.

Research on endocrine disruptors and epigenetics is ongoing and provides information about how toxic chemicals exert their effects: they are mechanisms of action. By revealing the likely biological pathways by which toxic substances cause effects on health, endocrine disruption and epigenetics make the scientific case against toxics even more credible and convincing. Given this fact, it's not surprising that the U.S. environmental health movement is using this new information in its work.

By the end of the twentieth century, the U.S. environmental health movement had become a vibrant social movement. With its roots in local and urban environmental health issues and grassroots activism, it transformed the debate about the environment from a scientific and legal discussion about federal policy into a national discussion about real people suffering from environmentally related diseases and disabilities. By asserting people's right to a healthy environment and working on locally identified issues such as toxic waste dumps, waste incineration, pesticides, and the right to know, the U.S. environmental health movement has achieved considerable success. It has also responded to the concerns of the American people in ways that the national mainstream environmental groups did not. In doing so, it built grassroots support and fostered local activism. Part II of this book picks up the story of the U.S. environmental health movement in the twenty-first century and examines the contemporary movement, including its organizations, issues, and strategies and how they contribute to social change.

NOTES

1. Adeline Gordon Levine. *Science, politics and people*. Lanham, MD: Rowman & Littlefield (1982).
2. Lois Marie Gibbs. *Love Canal and the birth of the environmental health movement*. 3rd edition. Washington, DC: Island Press (2010).
3. Lois Marie Gibbs. *Love Canal: The story continues*. Gabriola Island, BC: New Society Publ. (1998).
4. Lois Marie Gibbs. Learning from Love Canal: A twenty year retrospective. *Orion Afield*. (Spring 1998).
5. Eckardt C. Beck. *The Love Canal tragedy*. EPA Journal (1979).
6. University at Buffalo, State University of New York. University archives Love Canal collections. Available at: http://library.buffalo.edu/specialcollections/lovecanal/about. Accessed October 31, 2012.
7. Costle presses for immediate passage of Superfund, EPA press release September 11, 1980. Available at http://www.epa.gov/history/topics/cercla/03.html. Accessed October 31, 2012.
8. Office of Technology Assessment. *Coming clean: Superfund's problems can be solved*. Washington, DC: Office of Technology Assessment (1989).
9. National Academy of Sciences. *Environmental epidemiology, Volume 1: Public health and hazardous wastes*. Washington, DC (1991).
10. Peter Montague. After twelve years studying toxic dumps, government knowledge remains sketchy. *Rachel's Hazardous Waste News* #272 February 12, 1992. Available at: http://www.ejnet.org/rachel/rhwn272.htm. Accessed October 31, 2012.
11. As quoted in: Sanford Lewis, Brain Keating, and Dick Russell. Inconclusive by design: Waste, fraud and abuse in federal environmental health research. Environmental Health Network and National Toxics Campaign Fund 1992. Available at: http://www.ejnet.org/toxics/inconclusive.pdf. Accessed on October 31, 2012.
12. Devra Davis. *The secret history of the war on cancer*. New York: Basic Books (2007).
13. Julian B. Andelman and Dwight W. Underhill. *Health Effects from Hazardous Waste Sites* . Chelsea, MI: Lewis Publ. (1987).
14. Gary Cohen, John O'Connor, and Barry Commoner. *Fighting toxics: A manual for protecting your family, community and workplace*. Washington, DC: Island Press (1990).
15. Robert D. Bullard. Solid waste sites and the black Houston community. *Sociological Inquiry* 53: 273–288 (Spring 1983).
16. U.S. General Accounting Office. *Siting of hazardous waste landfills and their correlation with racial and economic status of surrounding communities*. Washington, DC: Government Printing Office. (1983).
17. Commission for Racial Justice. *Toxic wastes and race in the United States*. New York: United Church of Christ (1987).
18. Robert D. Bullard, Paul Mohai, Robin Saha, and Beverly Wright. *Toxic wastes and race at twenty: A report prepared for the United Church of Christ Justice & Witness Ministries*. (2007). Available at: http://www.ucc.org/assets/pdfs/toxic20.pdf. Accessed October 31, 2012.
19. Robert Bullard. *Dumping in Dixie: Race, class and environmental quality*. Boulder, CO: Westview Press (1990).

20. Cerrell Associates Inc. *Waste to energy, political difficulties facing waste-to-energy conversion plant siting.* Sacramento: California Waste Management Board Technical Information Series (1984).

21. D. J. Farren. *Report on the site selection process for the North Carolina "Low-Level" Radioactive Waste Disposal Facility.* Raleigh, NC: Chatham County Board of Commissioners and the Chatham County Site Designation Review Committee (1992).

22. Lois Gibbs. Citizen activism for environmental health: The growth of a powerful new grassroots health movement. *The Annals of the American Academy of Political and Social Science* 584: 97–109 (2002).

23. Doug Brugge, Jamie L. deLemos, and Cat Bui, The Sequoyah Corporation Fuels release and the Church Rock spill: Unpublicized nuclear releases in American Indian Communities. *American Journal of Public Health* 97(9):1595–1600 (2007).

24. Winona LaDuke. *All our relations: Native struggles for land and life.* Cambridge, MA: South End Press (1999).

25. Winona LaDuke. *Recovering the sacred: The power of naming and claiming.* Cambridge, MA: South End Press (2005).

26. Charles Lee. *Proceedings: The First National People of Color Environmental Leadership Summit.* New York: United Church of Christ Commission for Racial Justice (1992).

27. Principles of Environmental Justice. First National People of Color Environmental Leadership Summit held on October 24–27, 1991, in Washington, DC. Available at: http://www.ejnet.org/ej/principles.html. Accessed October 31, 2012.

28. Marianne Lavalle and Marcia Coyle. Unequal protection: The racial divide in environmental law. *National Law Journal* 21: S1–12 (September 1992).

29. Environmental Protection Agency. *Environmental equity: Reducing risk for all communities.* (June 1992). Available at: http://www.p2pays.org/ref/32/31476.pdf. Accessed October 31, 2012.

30. Release of environmental equity report, EPA press release, July 22, 1992. Available at: http://www.epa.gov/history/topics/justice/01.html. Accessed October 31, 2012.

31. *Executive Order 12898: Federal actions to address environmental justice in minority populations and low income populations.* Available at: http://www.epa.gov/Region2/ej/exec_order_12898.pdf. Accessed October 31, 2012.

32. Memorandum of Understanding on Environmental Justice and Executive Order 12898. Available at: http://www.epa.gov/environmentaljustice/resources/publications/interagency/ej-mou-2011-08.pdf. Accessed October 31, 2012.

33. Milton Kotelchuck and Gerald Parker. *Woburn Health Data Analysis 1969 – 1978.* Massachusetts Department of Health (December 21, 1979).

34. Gerald Parker and Sharon Rosen. *Woburn: Cancer incidence and environmental hazards 1969–1978.* Massachusetts Department of Health. (January 23, 1981).

35. Jonathan Harr. *A civil action.* New York: Random House (1995).

36. Marilyn Leistner. *The Times Beach story.* Proceedings of the Third Annual Hazardous Materials Management Conference, Philadelphia, PA, June 4–6, 1985.

37. *Great Lakes Water Quality* Agreement. International Joint Commission. (1978). Available at: http://www.epa.gov/glnpo/glwqa/1978/index.html. Accessed October 31, 2012.

38. For example: Greta G. Fein, Joseph L. Jacobson, Sandra W. Jacobson, Pamela M. Schwartz, and Jeffrey K. Dowler. Prenatal exposure to polychlorinated biphenyls: effects on birth size and gestational age. *Journal of Pediatrics* 105(2): 315–20 (1984).

39. *Great Lakes Water Quality Agreement.* International Joint Commission. (1978). Available at: http://www.epa.gov/glnpo/glwqa/1978/index.html. Accessed October, 2012.

40. Lee Botts and Paul Muldoon. *Evolution of the Great Lakes Water Quality Agreement.* East Lansing, MI: Michigan University Press (2005).

41. International Joint Commission. *Sixth biennial report under the Great Lakes Water Quality Agreement* (1992). Available at: http://www.ijc.org/php/publications/html/6bre.html#summary. Accessed October 31, 2012.

42. A breakthrough in control of toxics. *Rachel's Hazardous Waste News Issue* 284 (May 6, 1992). Available at: http://www.ejnet.org/rachel/rhwn284.htm. Accessed October 31, 2012.

43. Peter Montague. The recent history of solid waste: Good alternatives are now available. *Rachel's Hazardous Waste News Issue* 289 (June 10, 1992).
44. U.S. Environmental Protection Agency. *Municipal solid waste generation, recycling and disposal in the United States. Facts and figures for 2007.* Available at: http://www.epa.gov/osw/nonhaz/municipal/pubs/msw07-fs.pdf. Accessed October 31, 2012.
45. EPA. *Pesticides industry sales and usage: 2006 and 2007 market estimates.* Washington, DC (2011).
46. Robert van den Bosch. *The pesticide conspiracy.* Berkeley: University of California Press (1978).
47. David Weir and Mark Schapiro. *Circle of poison: Pesticides and people in a hungry world.* Oakland, CA: Food First (1981).
48. Natural Resources Defense Council. *Intolerable risk: Pesticides in our children's food.* Washington, DC (1989).
49. Roni A. Neff and Lynn R. Goldman. Regulatory parallels to Daubert: Stakeholder influence, "sound science," and the delayed adoption of health-protective standards. *American Journal of Public Health* 95 (Suppl. 1) : S81–91 (2005).
50. National Research Council Committee on Pesticides in the Diets of Infants and Children. *Pesticides in the diets of infants and children.* Washington, DC: National Academies Press (1993).
51. Richard Wiles, Kenneth A. Cook, Todd Hettenbach, and Christopher Campbell. *How 'bout them apples? Pesticides in children's food ten years after Alar.* Washington, DC: Environmental Working Group (1999). Available at: http://www.ewg.org/files/apples.pdf. Accessed October 31, 2012.
52. Rick Engler. How we won the 'right to know' in New Jersey. *Labor Notes.* Issue 56. p. 1, 15. September 27, 1983.
53. Ingrid Eckerman. The Bhopal saga—causes and consequences of the world's largest industrial disaster . India: Universities Press (2004).
54. EPA. Toxics release inventory: National analysis overview. (2011). Available at: http://www.epa.gov/tri/tridata/tri1/nationalanalysis/overview/2011_TRI_NA_Overview.pdf. Accessed October 31, 2012.
55. A legislative history of the *Superfund Amendments and Reauthorization Act* of 1986. Committee on Environment and Public Works, U.S. Senate, October, 1990, Volume 2, p. 1084.
56. Sandra Steingraber. *Living downstream: A scientist's personal investigation of cancer and the environment.* 2nd edition. Cambridge, MA: Da Capo Press (2010).
57. Massachusetts Coalition for Occupational Safety and Health. *Labor and toxics use reduction: Opportunities and challenges* (1996).
58. Vitruvius. *De architectura* Book 8, Chapter 6. (c. 15 BCE). Available at: http://penelope.uchicago.edu/Thayer/E/Roman/Texts/Vitruvius/8*.html. Accessed on October 31, 2012.
59. As quoted in Gerald Markowitz and David Rosner. *Deceit and denial: The deadly politics of industrial pollution.* Berkeley: University of California Press (2002). p. 14.
60. Gerald Markowitz and David Rosner. *Deceit and denial: The deadly politics of industrial pollution.* Berkeley: University of California Press (2002).
61. As quoted in: Gerald Markowitz and David Rosner. *Deceit and denial: The deadly politics of industrial pollution.* Berkeley: University of California Press (2002). p. 64.
62. Walter Troesken. *The great lead water pipe disaster.* Cambridge, MA: MIT Press (2006).
63. Jack Lewis. Lead poisoning: A historical perspective. *EPA Journal.* (May 1985).
64. Jack Lewis. Lead poisoning: A historical perspective. *EPA Journal.* (May 1985).
65. U.S. Environmental Protection Agency. *EPA's position on the health implications of airborne lead.* Washington, DC (1973).
66. Herbert L. Needleman, Charles Gunnoe, Alan Leviton, Robert Reed, Henry Peresie, Cornelius Maher, and Peter Barrett. Deficits in psychologic and classroom performance of children with elevated dentine lead levels. *New England Journal of Medicine* 300: 689–95 (1979).

67. Herbert Needleman and David Bellinger. Developmental Consequences of Childhood Exposures to Lead. In: *Advances in Clinical Child Psychology* Volume 7 edited by B. Lahey and A. Kazdin. New York: Plenum Press (1984)

68. Herbert L. Needleman, Julie Reiss, Michael J. Tobin, Gretchen E. Biesecker, and Joel B. Greenhouse. Bone lead levels and delinquent behavior. *Journal of the American Medical Association* 275(5): 363–369 (1996).

69. Blood lead levels—United States, 1999–2002. *Morbidity and Mortality Weekly Report.* Centers for Disease Control and Prevention 54(20): 513–516 (May 27 2005).

70. Preventing lead poisoning in young children. U.S. Department of Health and Human Services, Public Health Service, Centers for Disease Control (1991). Available at: http://wonder.cdc.gov/wonder/Prevguid/p0000029/p0000029.asp. Accessed October 31, 2012.

71. CDC Response to Advisory Committee on Childhood Lead Poisoning Prevention. *Low level lead exposure harms children: A renewed call for primary prevention.* Available at: http://www.cdc.gov/nceh/lead/ACCLPP/CDC_Response_Lead_Exposure_Recs.pdf. Accessed October 31, 2012.

72. Theo Colborn, Fred vom Saal, and Anna Soto. Developmental effects of endocrine-disrupting chemicals in fish and wildlife. *Environmental Health Perspectives* 101 (5): 378–84 (1993).

73. Theo Colborn, Dianne Dumanoski, and John Peterson Myers. *Our stolen future*. New York: Dutton (1996).

74. Andrea Baccarelli and Valentina Bollati. Epigenetics and environmental chemicals. *Current Opinions in Pediatrics* 21(2): 243–251 (2009).

Part II

The Contemporary Movement

Chapter Five

Organizations and Issues

Part II of this book examines the contemporary U.S. environmental health movement. Building on the historical and cultural roots outlined in Part I, the next five chapters describe how the movement has evolved in the twenty-first century. By examining its organizations, issues, and strategies today, Part II reveals how the U.S. environmental health movement is advancing social change.

The organizations and people working for environmental health have changed over the years. In the early days, local groups played a leading role. Principally concerned about the right to a healthy environment, they focused on local toxic chemical issues, grassroots organizing, and collective action. But by the early 2000s, state groups became more prominent, and unlike local groups, they emphasized lobbying for new legislation on toxics. Even though they worked at the state and local levels, the lobbying strategies used by state groups were similar to those used by the national mainstream environmental groups in the late 1960s and 1970s.

As a result, a tension emerged between state and local groups, at least in the minds of some local activists. While many local environmental health and justice groups believed that grassroots action on local issues was the most effective way to deal with toxic chemicals, many state groups preferred political negotiation and compromise. Although both opposed toxics and sought to protect environmental health, they went about it in different ways; their strategies for social change were different. Furthermore, state groups realized that they needed to coordinate their efforts and began to form national coalitions among themselves, as well as with diverse constituencies. To add to this, the need to strengthen the movement's scientific capacity and its access to information became clear, so national organizations and services were created to do this.

This chapter describes the local, state, and national organizations that comprise the U.S. environmental health movement and their work on toxic chemicals, as well as the roles of women's organizations, labor, and Judeo-Christian religions. It also examines how the movement is broadening its focus beyond toxic chemicals and working on other issues, including nanotechnology, electromagnetic fields, fossil fuels, and green building. The chapter concludes with a discussion of the significance of foundation funding for the movement.

THE MOVEMENT'S STRONGEST ASSET: STATE AND LOCAL GROUPS

Today, the U.S. environmental health movement's strongest asset is its state and local groups, unlike its early days when local groups predominated. In fact, there are now environmental health groups in every city and state across the entire country. Leading organizations include Alaska Community Action on Toxics; the Alliance for a Clean and Healthy Maine; Californians for a Healthy and Green Economy (CHANGE); the Center for Environmental Health in California and New York; the Citizens' Environmental Coalition in New York; Clean and Healthy New York; Clean Production Action in Massachusetts; Communities for a Better Environment in California; the Coalition for a Safe and Healthy Connecticut; the Connecticut Coalition for Environmental Justice; the Ecology Center in Michigan; the Environmental Health Strategy Center in Maine; the Institute for Agriculture and Trade Policy in Minnesota; the JustGreen Partnership in New York; the Just Transition Alliance in California; the Oregon Environmental Council; Toxic Free North Carolina; the Washington Toxics Coalition; and West Harlem Environmental Action (WE ACT). Even from this incomplete list, it's obvious that many groups are located in the Northeast and Midwest, and on the West Coast. This is because these regions are relatively progressive and tend to be more concerned about environmental health. However, it's also true that there are many groups in the southern states, where much of the U.S. chemical industry is located and where many poor and disadvantaged communities suffer disproportionate environmental health risks.

The movement's state and local groups are currently its strongest asset for several reasons. With their focus on environmental justice, local groups have always been its bedrock and have been successful at cleaning up toxic waste sites, closing incinerators, securing the right to know, and winning other victories. Indeed, no social movement can be successful without staunch and unwavering support from local grassroots groups. But since the early 2000s, their strength has been supplemented by state groups. By lobbying for new legislation and building national coalitions, they became successful at passing state and local legislation. This combination of strategies and types of groups is helping to advance social change.

The movement's decision to work on legislative reform at the state and local levels was very deliberate. Starting in the 1980s, a series of Republican administrations hostile to environmental protection and Congressional bickering made further progress on environmental protection at the federal level all but impossible. So by the early 2000s, state groups were targeting states and municipalities known to be sympathetic to environmental health, such as California, Massachusetts, and San Francisco. The strategy worked and resulted in many victories. In this way, state groups leveraged the shortcomings of the federal government into innovative state and local policies.

In total, over nine hundred state and local toxics policies were proposed or enacted between 1990 and 2009,[1] and between 2003 and 2011, eighteen states passed seventy-one chemical safety laws.[2] Paradoxically, this state and local legislation may actually pave the way for action at the federal level. By pilot testing different approaches, it provides concrete evidence about what works and what doesn't that can be used to lobby federal legislators.

Many state and local policies take aim at products and individual toxic substances, such as lead, mercury, and PBDEs. In recent years, state groups have been encouraging legislators to move away from a chemical-by-chemical approach and towards more comprehensive management strategies that address multiple chemicals at the same time. Because it can take years to pass a single law or regulation and because there are well over eighty thousand chemicals in use in the United States today, advocating for broader approaches makes a lot of sense. Different states have developed different types of comprehensive management approaches, but the main ones include policies to ban or control persistent bioaccumulative toxics (PBTs); policies on specific categories of products, such as children's products; and policies that establish frameworks for prioritizing substances for legislative action.

Not only have state groups been successful at passing legislation, they've also shortened the length of time it takes from when an issue is first raised on the political agenda to when action is taken. Because they've been so effective at increasing public awareness, it's difficult for legislators to ignore environmental health problems for long. For instance, it took many decades for lead to be banned from gasoline, but it's taking much less time to pass legislation on bisphenol A, PBDEs, and phthalates. Even so, most movement supporters would agree that legislative reform still takes too long.

According to one analysis of the state chemical policy campaigns, their success can be attributed to several factors including:

- "A health frame—the campaigns were sharply framed around children's health, not the environment, as well as the health of key constituencies (e.g., women and workers);

- Strong coalitions—diverse health-based coalitions were organized with capacity to apply targeted grassroots power, direct legislative advocacy, and strategic communications;
- A product focus—parents and policymakers easily related to chemical threats in the home from consumer products, which were less politically threatening to in-state industries;
- A split-the-opposition strategy—the out-of-state chemical companies and their allied national trade groups remained villains, not local businesses and green chemistry entrepreneurs; and
- Bipartisan wins—a series of winning campaigns built confidence and a bipartisan consensus that protecting children's health from the chemical industry was good politics."[3]

But despite the success of state groups at passing legislation on products and toxic chemicals, local groups often have a difficult time dealing with a different type of legislation. In particular, it's very difficult to prevent the siting and operation of hazardous facilities because federal and state regulatory systems are based on giving permits and approvals. If local groups follow federal and state rules, they can only argue that the necessary permitting processes have not been followed, that federal or state regulators have not considered all the relevant scientific factors, or that permit violations have occurred after the fact. Because of the permissive way that the regulatory system is structured, it's almost impossible to stop corporations from polluting the environment and damaging human health. So even though local activists may occasionally win a court battle, the system is stacked against them.

Recognizing this, some local groups are using a new approach. Rather than following federal and state rules, they're working to transform the entire regulatory system. By successfully lobbying for ordinances, home rule charters, and other local legislative instruments, they're creating environmental health protection systems that go beyond regulating harm and actually prevent it. For example, Blaine Township in Pennsylvania passed ordinances in 2006, 2007, and 2008 that ban coal mining, require businesses to publicly disclose their activities, and assert that corporations do not have constitutional rights as persons. Similarly, the cities of Santa Monica and Pittsburgh have adopted community bills of rights that provide their citizens with the right to a healthy environment that supersedes the rights of corporations. In fact, as of 2009, more than a hundred U.S. communities, with a total of about 350,000 residents, in at least four states had enacted ordinances, charters, and other measures asserting local democratic control over corporations and their environmentally damaging activities.[4] Pioneered by the Community Environmental Legal Defense Fund, these new legislative instruments allow municipalities to protect the environmental health of their citizens.

Anyone looking at the numbers and types of new state and local policies and ordinances on toxics might conclude that the legislative battle has been won, but appearances can be deceptive. Even though many states and municipalities have enacted protective policies, others haven't. Just as some strive to be national leaders, others have moved in the opposite direction, doing as little as possible to protect their citizens. The result is an uneven national patchwork, with some people receiving less environmental health protection than others. This situation is somewhat similar to the situation in the late 1960s and early 1970s when water and air pollution were still state or local responsibilities and the federal government stepped in to create a national environmental protection framework (see chapter 3).

Like then, people with fewer environmental health safeguards often live in areas with the greatest risks. For instance, despite the fact that the southern states are home to much of the U.S. chemical industry, many governments in the region are reluctant to enact strong legislation because it's seen as an obstacle to continued economic growth. As a result, residents of states such as Mississippi, Alabama, Louisiana, and Tennessee face a double whammy. Not only do they live with greater exposures, they have fewer legal protections. This situation creates environmental injustice that only the federal government has the power to fix.

Federal legislation is also needed because contaminants don't respect state or local boundaries any more than they respect national ones. Chemicals don't recognize political or geographic borders. Transported by air, water, birds, animals, and humans, they don't carry passports or visas. By its very nature, chemical pollution is a transboundary problem, and only federal legislation can deal with it. However, getting new federal legislation is proving to be very challenging and will necessitate even more concerted action by the U.S. environmental health movement, especially its national organizations and coalitions.

THE ROLES OF NATIONAL ORGANIZATIONS

At a national level, there are some groups working exclusively on environmental health, as well as environmental groups that use health-based arguments. Starting in the 1980s, when Ellen Silbergeld, the Environmental Defense Fund's senior toxicologist, launched a campaign on the health hazards of leaded paint, the environmental movement has steadily expanded its work on urban environmental health issues. Today, national environmental groups, such as the Sierra Club, the Natural Resources Defense Council, and the Environmental Defense Fund, often use health-based arguments.

National organizations have many essential roles in the U.S. environmental health movement, including developing big-picture strategies and ideas, raising public awareness and providing information to the public, translating scientific information so that it can be more easily understood, providing support to local groups, and perhaps most importantly, reaching out to diverse constituencies and coordinating all the different groups and constituencies that make up the movement today.

As the U.S. environmental health movement matures, it's critically important to develop big-picture strategies and ideas. This is where organizations such as the Environmental Health Fund, Commonweal, and the Science and Environmental Health Network (SEHN) excel. The Environmental Health Fund was founded in October 1998 as a coordinating and fundraising organization, and over the past fifteen years, it's functioned as a central architect for the movement's advocacy efforts. By guiding the development of new organizations, coalitions, and campaigns, such as Health Care Without Harm (HCWH), SAFER States, Safer Chemicals, Healthy Families, and many others, it has quietly helped to mastermind the movement's growth. Another organization that's played a significant role in growing the movement is Commonweal, a nonprofit health and environmental research institute founded in 1976 by Michael Lerner. With programs on human and ecosystem health, it's given birth to many environmental health organizations, including HCWH, the Collaborative on Health and the Environment (CHE), and the Health and Environment Funders Network (HEFN). SEHN is also a big-picture thinker but in a very different way. Established in 1993, SEHN is a "think tank" that has played a key role in formulating and promoting precautionary approaches (see chapter 7), policies to protect the environmental commons for future generations, and measures to reduce cumulative impacts. Its innovative and forward-looking ideas make it one of the movement's most influential thought-leaders.

A second role for national organizations is raising awareness about environmental health issues by providing information and resources to the public. Indeed, this is one of the U.S. environmental health movement's most important functions. National as well as state groups have become "go-to" places for people looking for information. Hungry for knowledge about how environmental quality affects health and how to reduce their exposures, people pick up the phone or send texts and e-mails. They want information on alternatives to pesticides, safe cosmetics, green household cleaning products, nontoxic sunscreens, how to get rid of unwanted pharmaceuticals, and hundreds of other topics. Because many people don't trust government agencies or corporations to provide them with the facts, the public often turns to environmental health groups for answers. By responding to this demand, they can raise awareness and attract new supporters.

One of the national groups that provides information to the public is the Environmental Working Group (EWG). Founded in 1993 by Ken Cook and Richard Wiles, this organization has an especially hard-hitting investigative style. By conducting groundbreaking research and analyzing information in obscure government databases, EWG has come up with some extremely powerful facts and figures which it has assembled into resources such as the *Shoppers' Guide to Pesticides.*[5] EWG is also marked by its independence. While many groups collaborate with each other, EWG often goes its own way and develops its own advocacy positions. With its excellent research, public education, and lobbying skills, the EWG is one of the most influential environmental health groups in the United States.

A third critically important national function is translating the torrent of increasingly complex scientific information into language that can be more easily understood. Since 2002, the Collaborative on Health and the Environment (CHE) has been one of the organizations with this expertise. Cofounded by Michael Lerner, president of Commonweal, CHE is a national and international partnership of over forty-five hundred individuals and organizations in almost eighty countries and all fifty states, including scientists, health professionals, people affected by environmentally related diseases, nongovernmental organizations, and concerned citizens. Its mission is to strengthen the scientific dialogue on the environmental factors that impact human health and facilitate collaborative efforts on environmental health.[6] Through its working groups, scientific databases, conference calls, and numerous meetings and events, CHE helps improve the movement's scientific literacy, which in turn helps it to make more effective arguments for protecting environmental health.

A fourth role is providing support, resources, and advice to local environmental health groups. One of the oldest and best-known groups that fulfills this function is the Center for Health, Environment and Justice (CHEJ), founded by Lois Gibbs in 1981 and originally called the Citizens' Clearinghouse for Hazardous Waste (see chapter 4). Through training, coalition building, and one-on-one technical and organizing assistance, CHEJ works to level the playing field so that ordinary people can have a say in the environmental policies and decisions that affect their health and well-being. By building political pressure from the bottom up, it works towards systemic change at the national level. Whether it's blocking a proposed asphalt plant, stopping the expansion of a waste transfer station, or fighting air pollution from a freeway, CHEJ empowers neighborhood groups by providing them with the tools they need to build a movement, conduct research, negotiate, and win.

But perhaps the single, most important function of national environmental health groups is to reach out to diverse constituencies and coordinate all the different groups and interests that make up the U.S. movement today. This is another of CHE's strengths. By intentionally courting people with environmentally related diseases and disabilities and the organizations that represent them, it's broadened the movement's base of support. And as the movement has grown and diversified, effective coordination has become more important. Today, most movement coordination is carried out by national coalitions, and one of the first national coordinating coalitions was SAFER States, formed in 2005. Providing much-needed forums for collaborative planning and strategizing, as well as platforms for raising public awareness, national coalitions are now an essential feature of the U.S. environmental health movement. But before exploring the role of national coalitions, it's important to consider how the reform of European toxics legislation influenced the U.S. environmental health movement and the emergence of national coalitions.

THE INFLUENCE OF EUROPEAN TOXICS POLICY

In the late 1990s, when U.S. groups were organizing to lobby for stronger state and local policies, European environmental health groups were fighting their own battles. In 1998, their work to highlight the weaknesses of European toxics legislation led to a European Commission study,[7] which, in turn led to a 1999 conference and to a widespread recognition that a new approach to chemicals management was urgently needed.[8] Encouraged by this success, European environmental health groups met in Copenhagen to formulate a common advocacy position. The resulting *Copenhagen Chemicals Charter* (2000) contained five demands:

1. A full right to know—including what chemicals are present in products.
2. A deadline by which all chemicals on the market must have had their safety independently assessed. All uses of a chemical should be approved and should be demonstrated to be safe beyond reasonable doubt.
3. A phase out of persistent or bioaccumulative chemicals.
4. A requirement to substitute less safe chemicals with safer alternatives.
5. A commitment to stop all releases to the environment of hazardous substances by 2020.[9]

In early 2001, the European Commission announced its proposed new policy.[10] An intense and acrimonious debate between government agencies, nongovernmental organizations, and the chemical industry followed. Using the demands of the *Copenhagen Chemicals Charter*, nongovernmental organizations lobbied hard. Key among them were the World Wide Fund for Nature (WWF), which launched a campaign called DetoX,[11] the International Chemical Secretariat (ChemSec), Greenpeace, the Health and Environment Alliance (HEAL), the European Public Health Alliance (EPHA)-Environment Network (EEN), and many labor, consumer, and animal rights organizations. Although most U.S. groups could only support their colleagues from afar, the European Environment Bureau organized several meetings that included U.S. participants, and in 2003 the Lowell Center for Sustainable Production organized a tour for European chemicals policy experts to stimulate discussion in the United States.

After five years and many compromises, REACH (Registration, Evaluation, Authorization and Restriction of Chemical Substances) was finally approved. It entered into force in 2007 in all of the twenty-seven countries comprising the European Union. Although there have been many problems with the way in which REACH has been implemented, it is probably the most progressive toxic chemicals management system in the world today and far surpasses the U.S. Toxic Substances Control Act (1976). Its features include:

- A precautionary approach.
- Requirements for chemical companies to provide safety data on the substances they manufacture and import.
- Provisions for the substitution of the most dangerous chemicals with safer alternatives.
- Removal of the artificial distinction between new and existing substances inherent in most chemical management systems.

As well as providing Europe with a better toxics management system, REACH had several effects in the United States. First, it highlighted the need to reform the antiquated *Toxics Substances Control Act* (1976). By providing a more up-to-date and effective approach, REACH shone a spotlight on the inadequate management system in the United States. Second, it imposed new, more rigorous safety requirements on U.S. chemical manufacturers who wanted to export their products to Europe, because they had to comply with them to retain their export markets. And third, it helped to reinvigorate U.S. environmental health groups at a bleak time in federal politics. Drawing strength from their colleagues' success, U.S. groups witnessed the benefits of developing a common advocacy platform and strong, sustained collaboration. So, even as negotiations on REACH were unfolding, U.S. groups worked together to develop a shared set of principles known as the *Louisville Charter*.

THE *LOUISVILLE CHARTER*

In May 2004, representatives of leading U.S. environmental health, community, labor, public health, and environmental organizations came together in Louisville, Kentucky, to discuss a new national strategy for chemicals policy reform. Although some groups had already been holding regular meetings, this time the purpose was much more strategic—the development of a shared national action agenda for legislative reform. Assisted by their earlier commitment to the 1998 *Wingspread Statement on the Precautionary Principle* (see chapter 7), and the *Copenhagen Chemicals Charter*, they soon reached agreement. Released at the conclusion of the meeting, the *Louisville Charter for Safer Chemicals*[12] called for fundamental reform of toxic chemicals management in the United States. It advocated:

- Altering production processes, substituting safer chemicals, redesigning products and systems, rewarding innovation and reexamining product function;
- Phasing out persistent, bioaccumulative, or highly toxic chemicals;
- Giving the public and workers the right to know;
- Acting on early warnings and preventing harm from new or existing chemicals when credible evidence of damage exists—the precautionary approach;
- Requiring comprehensive safety data for all chemicals, also known as the principle of "no data, no market"; and
- Taking immediate action to protect communities and workers, so that no population is disproportionately burdened by chemicals.

Hailed as "the NGO blueprint for new chemicals policy,"[13] the *Charter* was a major step forward. For the first time, U.S. groups had a common understanding of what needed to be done. Using the *Charter* as a unifying foundation, state groups subsequently developed their own campaigns and programs to lobby for legislative reform.

As well as representing a national consensus, the *Louisville Charter* marked an important shift in thinking. From focusing on reducing, preventing, and cleaning up toxic chemicals, the U.S. environmental health movement began to stress the need for safer substances. This more positive approach laid the groundwork for later work on green chemistry and safer materials (see chapter 9).

THE EMERGENCE OF NATIONAL COALITIONS

U.S. environmental health groups had started to collaborate long before the Louisville meeting, and in 2001, they launched a national media campaign to promote Bill Moyers' documentary on the chemical industry, *Trade Secrets*. This led to the formation of the national Coming Clean collaborative, which provides a movement-wide forum for information sharing and strategy development. Its Policy Work Group adopted the *Louisville Charter* and, in 2005, helped to create SAFER States (State Alliance for Federal Reform of Chemicals Policy) to coordinate the movement's efforts to strengthen toxic chemicals legislation. A coalition of environmental health groups in states that have demonstrated leadership in toxics policy reform, including California, Maine, Massachusetts, Michigan, Minnesota, New York, Oregon, and Washington, SAFER States' vision declares: "We believe families, communities, and the environment should be protected from the devastating impacts of our society's heavy use of chemicals. We believe that new state and national chemical policies will contribute to the formation of a cleaner, greener economy."[14]

The state organizations that make up SAFER States have done a lot of work together. In 2005, they supported Senator Frank Lautenberg (D) when he introduced the *Kid Safe Chemical Act* into the Senate and Representative Henry Waxman's (D) companion bill in the House. Intended to reform the *Toxic Substances Control Act* (1976), both bills died. Undaunted, SAFER has gone on to plan and execute state campaigns for legislative reform across the entire country. Groups share information, organize concurrent events, and learn from each other's successes and failures. For instance, in 2013 groups in twenty-six states worked together to propose toxics legislation in their own jurisdictions.[15]

With the arrival of the Obama administration in 2009, the prospects for new federal legislation on toxic chemicals improved significantly. Even as President Obama was swearing the oath of office, U.S. environmental health groups boosted their work for stronger toxics legislation by launching a new national coalition. Called *Safer Chemicals, Healthy Families*, it built on the success of SAFER States but included diverse organizations and individuals, including parents, health professionals, advocates for people suffering from environmental health diseases, labor organizations, faith groups, environmentalists, businesses, and environmental justice activists, as well as national, state, and local environmental health groups. Its platform includes:[16]

- Taking action on the most dangerous chemicals; phasing out persistent, bioaccumulative toxics; expanding green chemistry research; and encouraging the use of safer chemicals;
- Holding industry responsible for the safety of its chemicals and products by requiring chemical companies to provide full information on the impact of their chemicals on health and the environment; and
- Using the best science to protect all people and vulnerable groups and encouraging studies on people's body burdens of toxic chemicals.

Agreement on this common position was relatively straightforward because some of the participating groups had previously worked together on the *Louisville Charter* or were members of SAFER States. Without this earlier collaborative work, it probably wouldn't have been possible to organize a campaign for federal toxics policy reform as quickly or as easily. In its first year, *Safer Chemicals, Healthy Families* grew rapidly, and by mid-2010, it comprised nearly 250 organizational members, representing more than eleven million people.

In April 2010, Senator Frank Lautenberg (D) introduced the *Safe Chemicals Act*[17] in the Senate, and Representatives Bobby Rush (D) and Henry A. Waxman (D) unveiled similar legislation in the House.[18] Intended to replace TSCA, these bills went a long way to meet *Safer Chemicals, Healthy Families*' platform and took account of Europe's REACH initiative. In particular, they would have required chemical manufacturers to prove the safety of their products and provide a minimum data set for each chemical. They would also have established a public database containing information submitted by chemical manufacturers and the EPA's risk assessments. However, with Democratic losses in the House of Representatives in the 2010 midterm elections, it became clear that new federal toxics legislation would not be passed quickly or easily. But TSCA reform is far from dead, and the U.S. environmental health movement continues to lobby for reform.

One of the strengths of the Safer Chemicals, Healthy Families Coalition is that it provides a forum for discussion among its diverse members. One of the main topics of debate is whether it should lobby for visionary, idealist legislation or be more pragmatic and go with what's politically winnable. But whatever their views, members are committed to continuing the coalition because they appreciate the benefits of national, multisectoral coalitions. By bringing together diverse constituencies, raising awareness, and demanding change, *Safer Chemicals, Healthy Families* is realizing the truth behind the old adage "united we stand, divided we fall" and playing a central role in keeping the issue of toxic chemicals on the federal government's agenda.

COMMUNICATIONS AND GETTING THE WORD OUT

The U.S. environmental health movement and its coalitions wouldn't be able to function unless they had the ability to communicate and get the word out. Indeed, effective communication is a prerequisite for collaborative action of any kind, and it's been made much easier by new electronic and digital tools. Because many U.S. environmental health groups came of age at the same time as the internet and the world wide web, these tools are second nature to them. By the 1990s, many groups had become early adopters of e-mail and launched web sites. This increased their public visibility and the capacity to attract new supporters, but by the early 2000s, it was obvious that the U.S. movement also needed a daily news service to keep it up-to-date. So in 2002, John Peterson Myers, coauthor of *Our Stolen Future*, established the Environmental Health News and started to publish a free daily e-newsletter called *Above the Fold*, which compiles U.S. and international news on environmental health. Today, *Above the Fold* is sent out to thousands of subscribers, and its syndication services are used by more than three hundred web sites around the world. Other services with environmental health news include the Environmental News Network and the Environmental News Service. By informing and educating the public, these services have helped to grow the U.S. environmental health movement.

Recognizing the potential of electronic media, many groups were also early adopters of Facebook, Twitter, You Tube, and other social media. Even though they don't have full-time social media staff, they use social media to communicate with the public, raise awareness, generate funding, and lobby legislators. As a consequence, the movement can now spread its message to anyone with a cell phone or a computer, organize flash mobs, and instantly create and distribute news stories and images to almost every media outlet on the planet. Time and geography are no longer constraints on communicating or organizing. What used to take months or years now takes only a few minutes.

The very first nongovernmental electronic database on environmental health was developed in the early 1980s by journalist Peter Montague. It provided local activists with free, up-to-date scientific data and information. Initially called the Hazardous Substances Database, and later renamed the Remote Access Chemical Hazards Electronic Library (RACHEL), it could be accessed through early telephone modems. A few years later, in 1986, Montague started to publish a weekly editorial newsletter called *Rachel's News*, which he later renamed *Rachel's Environment & Health News* and then *Rachel's Democracy & Health News*. Although it ceased publication in 2009, this newsletter provided the first regular source of information and commentary on environmental health and was a major influence on the movement's development.

THE IMPORTANCE OF WOMEN'S ORGANIZATIONS

The U.S. environmental health movement's ability to communicate and get the word out has been greatly enhanced by women. Since the growth of social media in the mid-2000s many women, especially stay-at-home moms, have started to blog about their lives and the issues that concern them, including environmental health. With names like Non-Toxic Kids, Enviromom, Mommy Greenest, and SafeMama, mom blogs have become an important way for many women to exchange information and talk about environmental health.

In fact, women play a disproportionately large role in the U.S. environmental health movement. The majority of national, state, and local groups are led by women, and there are several women's environmental health organizations. One of the most important is Women's Voices for the Earth (WVE). Founded in 1995, WVE is a national group, based in Missoula, Montana, working on a variety of issues including safe household cleaning products, cosmetics, healthy nail salons, and reducing mercury pollution. Another women's environmental health organization is EcoMom Alliance. With over thirty thousand followers, fans, and subscribers in its social media network, EcoMom reaches over one million families each year. Its campaigns include the EcoMom Alliance Challenge, Green Halloween, and Just Label It (a campaign advocating the labeling of genetically modified food). As well as these and other women's environmental health organizations, many women's organizations are active supporters of environmental health, including the League of Women's Voters, Planned Parenthood, Moms Rising, Code Pink, and SisterSong.

Today's women's environmental health groups are the descendants of several earlier initiatives, including Mothers and Others, founded in 1989 by actress Meryl Streep; the Philadelphia-based Women's Health and Environment Network, which merged with the Philadelphia chapter of Physicians for Social Responsibility; a series of conferences on Women's Health and the Environment started by Teresa Heinz in 1996; and the Women's Health and Environment Initiative, which grew out of the first Women's Health and the Environment conference and is now a project of Women's Voices for the Earth. These initiatives helped to blaze the trail for present-day organizations.

Women's participation and leadership in the U.S. environmental health movement are important because women have the power to make a significant difference. Ever since the suffrage movement of the early twentieth century, women have become increasingly politically active and visible in public life. Comprising more than 50 percent of the population, they're a social, cultural, and economic force to be reckoned with. But women are important for other reasons too. First, they face greater environmental health risks than men. They use more toxic chemicals, handle cleaning products more frequently,[19] and use twice as many personal care products a day as men on average (twelve as compared to six).[20] And because women have a higher percentage of body fat, where many toxics accumulate, they can have larger amounts. There's also evidence that these exposures translate into environmentally related diseases. Over the past twenty years, breast cancer rates have risen from a lifetime risk of one in twenty to one in eight[21] —an alarming increase by any standards. Several studies have shown a connection between environmental endocrine disruptors (see chapter 4) and breast cancer risk,[22, 23] and at least one has shown an increase in breast cancer rates linked with higher exposures to toxic chemicals.[24] To add to this, millions of women are affected by reproductive disorders associated with toxic chemicals, including early breast development, premature menarche, uterine fibroids, endometriosis, and polycystic ovarian syndrome.

A second reason why women's participation and leadership is important is that American women make approximately 80 percent of the health-care decisions for their families and are more likely to be caregivers than men.[25] As mothers, sisters, wives, grandmothers, aunts, daughters, and friends, women are often the ones to look after family members when they get sick and encourage others to protect their health.

But perhaps the most important reason is that many women are mothers. Women conceive, carry, and nourish their developing offspring inside their bodies for nine whole months. They give birth and nurse their newborn babies. And then they feed, clothe, and look after their children until they reach adulthood. And although some may dispute that this creates a unique bond, the reality is that women are passionate about protecting the health of their children and future generations.

Given these facts, it's natural for women and women's organizations to play a central role in the U.S. environmental health movement. This isn't to diminish or downplay men at all; they, too, play a role. One of the ways they do this is through labor organizations. Mostly led by men, labor organizations are forging new alliances with environmental and environmental health organizations.

ALLIANCES WITH LABOR ORGANIZATIONS

Although there was some limited collaboration between environmental groups and labor organizations in the 1980s and 1990s (see chapter 4), there weren't any long-term, national alliances. But by the late 1990s, the situation began to change. One of the first signs of a growing rapprochement was the development of the Just Transition framework. Originally proposed by labor unions and supported by environmental groups, this framework makes a case for sustainable production, social justice, green jobs, representation in decision making, and education and training. Since then, Just Transition has become an organizing principle that brings together workers, communities, and environmental justice groups to help create a sustainable, just economy. One of the leading organizations is the California-based Just Transition Alliance, which strives to reduce worker and community exposure to toxics and build fair, locally based economies. Founded in 1997, it is a regional coalition of labor, environmental justice activists, Indigenous people, and working-class people of color.

In the mid-2000s, the environmental movement and organized labor took another step together and established the national BlueGreen Alliance.[26] Launched in 2006 by the United Steelworkers and the Sierra Club, this partnership includes fourteen of the country's largest unions and environmental organizations. Acting together, through more than fifteen million members and supporters, the Alliance is a powerful voice for building a "green economy" (see chapter 9). For the first time in more than a century, activists working to improve the environment and jobs are collaborating at a national level, creating a forum that could be used to bridge the gap between environmental and occupational health.

The creation of the BlueGreen Alliance is a very welcome development because it signifies a deepening understanding about the connections between human health, the economy, and the environment. Since the economic recession of 2008 and the high unemployment rates that followed, environmentalists have begun to understand the importance of wages, job security, and other labor issues. At the same time, labor organizers are becoming aware that the current economic system is unsustainable over the long term and threatens human health. As a result, the two movements are now beginning to transcend their differences and work together more systematically and strategically.

National labor unions are important allies, even though they may be less crucial than in the past. Their falling membership rolls and reduced political influence means that they lack the power they once had, but their resources and expertise at grassroots organizing, institutional infrastructure, and national reach are still very impressive and could help to raise awareness about environmental health problems among millions of U.S. workers.

Given the declining power of national unions, local partnerships are becoming even more important. Growing out of the COSHs (Committees on Occupational Safety, and Health) established in the 1970s (see chapter 3), there are now many state, regional, and local coalitions on environmental and occupational health. Creating alliances between the factory floor and the grassroots, they comprise local activists concerned about occupational and environmental health, working conditions, social justice, and economic development. For instance, the Alliance for a Healthy Tomorrow in Massachusetts includes members from the local AFL-CIO, the Service Employees International Union (SEIU), and the Massachusetts Teachers' Association.

The new desire to build long-term partnerships between workers, organized labor, and environmentalists could benefit the U.S. environmental health movement and expand its base of support. In particular, it could help advance work on green chemistry and safer materials (see chapter 9). Collaborating with workers directly involved in industrial production will help to ensure that efforts to develop safer, nontoxic substances and redesign manufacturing processes are practical, relevant, and useful. Despite widespread automation and the outsourcing of thousands of U.S. jobs to other countries, American workers could play an important role in the struggle for environmental health. Their knowledge and experience could help to make industrial production less harmful to worker and community health. For these reasons, workers and organized labor are significant allies for the U.S. environmental health movement.

NEW WAYS OF FRAMING ENVIRONMENTAL HEALTH: JUDEO-CHRISTIAN RELIGIONS

At the same time as labor organizations are becoming concerned about environmental health, so are Judeo-Christian religions. By extending their teachings about social justice to include the environment, they're helping to frame environmental health issues in ethical, spiritual, and moral terms. For instance, drawing on the well-known Christian exhortation to "love your neighbor as yourself," one Judeo-Christian environmental group asks, "Can we love our neighbor without protecting the environment on which that neighbor's life and health depend?"[27] Although Judeo-Christian religions foster a belief in anthropocentrism—a primary cause of many environmental health problems (see chapter 1)—they may also be part of the solution. By combining a commitment to social justice with the idea of environmental stewardship, these religions could play an important role, although, like labor unions, their membership rolls are falling.

Since the late 1970s and 1980s when southern black churches protested against toxic waste dumps, Judeo-Christian religions have become increasingly active on environmental health. Today, they're working on many issues, including strengthening toxics legislation, children's environmental health, and water and air pollution. In 1993, Evangelical Christians established the Evangelical Environmental Network "to equip, inspire, disciple, and mobilize God's people in their effort to care for God's creation."[28] Together with the U.S. Conference of Catholic Bishops, this network supported stronger regulations on mercury and other chemicals that affect the "unborn child."[29,30] In 2009, the National Council of Churches (NCC) passed a resolution[31] stating "The National Council of Churches will work to protect the most vulnerable members of society from exposure to harmful chemicals—pregnant women, children, communities of color, low-income communities, older adults and others with compromised immune systems, and people exposed to these chemicals in the workplace." Campaigning alongside secular environmental health groups on toxics policy reform, cosmetics and women's and children's environmental health, the NCC has become a staunch movement ally. To add to these organizations, there are several supportive interfaith groups including the National Religious Partnership for the Environment, Earth Ministry, Voices for Earth Justice, and GreenFaith. In 2010, many national and state faith-based organizations released an "Interfaith Statement for Chemical Policy Reform" that calls for Congress to overhaul the *Toxic Substances Control Act* (1976).[32]

To add to this, some Judeo-Christian religions are active on climate disruption. In 2000, the Coalition on the Environment and Jewish Life released a statement called *Global Warming: A Jewish Response*,[33] and in 2006 the Evangelical Climate Initiative issued a *Call to Action*[34] and the Catholic Coalition on Climate Change was launched.[35] At the international level, religious leaders released the Interfaith Declaration on Climate Change[36] in 2009, as well as a Statement of the World's Religious Traditions on Climate Change.[37] Indeed, climate disruption is one of the few issues that most of the world's faith traditions can agree on.

The growing interest of Judeo-Christian religions in environmental health suggests they could play a larger role in the movement. Despite the fact that the number of Americans who do not identify with any religion continues to rise, they're important allies. Not only do religious groups broaden the environmental health movement's base of support, they also add new, critically important ways of framing the issues. Religious groups make it very clear that environmental health is a moral, spiritual, and ethical concern, not just a scientific, technical, or economic one. By emphasizing this, they can help to broaden and deepen the debate. Moral, spiritual, and ethical choices lie at the very heart of human life, so by framing environmental health issues in these terms, religious groups can add significantly to the public discourse. Moreover, recasting environmental health problems in moral, spiritual, and ethical ways reveals that they can't be solved through technical or engineering approaches alone and that fundamental changes in social values and beliefs are needed. To add to this, because most religious leaders are well-respected members of society, people sit up and take notice when they speak out on issues. They can do a lot to raise awareness and shape public opinion. When religious leaders talk about the need for environmental health, people listen. For all these reasons, increased participation of Judeo-Christian religions would likely benefit the U.S. environmental health movement.

So far, this chapter has discussed the movement's work on toxic chemicals and the organizations that do it. This is because this issue has dominated the movement's efforts for the past thirty-five years. Indeed, the U.S. environmental health movement is sometimes regarded as an antitoxics movement, but this discounts work on other issues. In fact, many national, state, and local organizations are now working on other issues, including nanotechnology, electromagnetic fields, fossil fuels, and green building. This situation has contributed to a lively debate within the U.S. environmental health movement: Should it stay focused on toxic chemicals, or should it expand it's scope? Whatever the outcome, some groups have taken a lead and are already working beyond toxics.

BEYOND TOXICS: NANOTECHNOLOGY

Consider nanotechnology. Since the early 2000s, a few U.S. groups, including the Natural Resources Defense Council, have been working on this issue. Troubled by how little is known about the health effects of nanotechnology and its rapid infiltration into consumer products, they have called for more research and tighter regulations.

Nanotechnology actually refers to all the different technologies used to manipulate materials at very small scales, including the manufacture of nanoparticles and nanotubules and products with minute design elements, such as water filters with microscopic pores. Used for many different purposes, nanomaterials are now in everything from sunscreen to household paint to anti-odor socks and glass cleaners. They are even present in foods and pharmaceuticals. In 2012, the Consumer Products Inventory of the Project on Emerging Nanotechnologies listed 1,317 products, produced by 587 companies, located in 30 countries.[38] This represents a more than sixfold increase since the inventory was first put together in 2006. At the same time, the markets for nanotechnology products are exploding. According to Lux Research, a market analysis firm, about $147 billion worth of nano-enabled products were sold in 2007, and this figure is expected to top $3.1 trillion by 2015.[39]

Despite the fact that nanomaterials are becoming commonplace, very little is known about their health effects, but what is known suggests they may be harmful. Nanomaterials tend to be more chemically reactive, more persistent, and more likely to pass through biological tissues. Many are so small they can enter the lungs, pass through cell membranes, and perhaps penetrate skin. Once inside the body, nanomaterials appear to have unlimited access to tissues, organs, and physiological systems, including the brain and possibly fetal circulation. Animal studies suggest that some can cause inflammation, damage brain cells, and lead to precancerous lesions. It's also known that ultrafine air pollution, much of which is nano-sized, is associated with reduced lung function and an increased likelihood of asthma, respiratory disease, and deaths from lung and heart disease. A review article in the prestigious journal *Environmental Health Perspectives* concluded: "There is reason to suspect that NPs (nanoparticles) . . . are likely to cause diseases—some with a long latency. With widespread industrialization of nanotechnology, there is the potential for ambient air pollution and a conceivable threat to the general population."[40]

With alarm bells like this sounding, one might assume there are special regulations in place. But there aren't. Nanoscale chemicals are managed using the same legislation as ordinary-sized substances. This doesn't make sense because the toxicological and physical properties of nanomaterials are different and can't be predicted from the properties of their normal-scale counterparts. In addition, it can be impossible to monitor levels of nanomaterials in the environment. For these reasons, environmental health activists argue that nanotechnology should be regulated separately. Although this hasn't happened yet, a research team funded by the National Institutes for Health has proposed draft guidelines for the oversight of nanomedicine.[41]

With no effective regulations in place, it's no wonder that environmental health groups are concerned. In 2007, a global coalition of more than forty organizations called for a comprehensive management approach,[42] and in the United States, the Natural Resources Defense Council proposed regulations based on prohibiting the use of untested or unsafe nanomaterials, requiring lifecycle assessments, and full and meaningful public participation in decision making.[43] A less stringent scheme was developed by the Environmental Defense Fund in partnership with chemical giant DuPont.[44]

BEYOND TOXICS: ELECTROMAGNETIC FIELDS

Another issue that has attracted the attention of some environmental health groups is electromagnetic radiation. Resulting from the production, transmission, and use of electric power and electronic equipment, including cell phones, electromagnetic fields (EMFs) and radio frequency (RF) have been linked with numerous health problems. Although many local groups have been concerned about this issue since the late 1970s, only a few national and state groups have weighed in.

Health concerns about EMFs were first raised in 1979 in a scientific paper which linked an increased risk of childhood leukemia with living close to power transmission lines.[45] Since then, more than a dozen studies have confirmed a connection between higher rates of childhood cancer and living near high voltage-lines. More recently, researchers have studied the effects of radiation from cell phone use and reported thermal heating, electromagnetic hypersensitivity, sleep disorders, brain cancers, and genotoxicity. Children appear to be more sensitive than adults. As well, research shows that people living close to cell phone towers can experience fatigue, headache, sleep disruption, and memory loss.

Despite the mounting evidence, government agencies continue to offer reassurance to the public. In 2013, the Food and Drug Administration (FDA) web site said that "available scientific evidence . . . shows no increased health risk due to radiofrequency (RF) energy, a form of electromagnetic radiation that is emitted by cell phones."[46] This statement puts the U.S. government at odds with international health agencies. In 2002, the International Agency for Research on Cancer (IARC) classified extremely low frequency electric and magnetic fields as "possibly carcinogenic to humans,"[47] a classification that was reconfirmed by the World Health Organization in 2007.[48] Also in 2007, an international group of scientists and public health professionals (the BioInitiative Working Group) released a report which concluded, "existing public safety limits are inadequate for both ELF (extremely low frequency fields) and RF (radiofrequency radiation)."[49] Then in 2010, a major decade-long international study—the Interfone Study—reported that people who use cell phones the most have a substantially elevated risk of brain tumors compared to people who do not use them.[50] Strangely, the same study concluded that cell phone users show no increased risk of developing brain tumors. So what's the truth?

The truth is that we don't know. But even if the risks of radiation from cell phone use are very small, the health consequences will be huge at a population level because almost everyone uses them. Because of this, some environmental health groups, including the Environmental Working Group, are recommending that people, especially children, limit their exposures. Internationally, several countries have already recommended a precautionary approach to cell phone use, including Germany, France, Austria, the UK, and Russia,[51] and in 2010 San Francisco became the first U.S. municipality to pass an ordinance requiring retailers to provide information on how much radiation is emitted by the cell phones they sell,[52] however, its implementation was blocked after corporate interests objected.

This issue, like nanotechnology, seems to be a rerun of the decades-long controversy regarding toxic chemicals. If regulators wait for scientific studies to completely prove causation (see chapter 7), many people could get sick or even die. Raising the question of how much scientific evidence is necessary to justify regulatory action, EMFs and nanotechnology would seem to be excellent examples of the need for precautionary measures.

BEYOND TOXICS: FOSSIL FUELS

In recent years, some environmental health activists have become concerned about the health effects of fossil fuel extraction and use. This isn't only about the health effects of climate disruption, it's also about the health implications of natural gas fracking and the health effects of burning coal. Fracking technology—the hydraulic fracturing of rock to release natural gas—has been around for decades, but now natural gas producers are using a new drilling method called "high volume horizontal hydraulic fracturing" to release gas locked in shale formations underneath Pennsylvania, Ohio, West Virginia,

Texas, Louisiana, Arkansas, Colorado, Wyoming, and other states. This technology has spread extremely rapidly, and the United States is now in the midst of an unprecedented gas drilling frenzy, second only to the gold rush of the nineteenth century. By 2012, there were about 493,000 wells across thirty-one states, almost double the number in 1990.[53]

But this goldmine of energy has been linked with numerous environmental health problems, including:

- The hydraulic fluid used to fracture the rock contains benzene, toluene, ethylbenzene, xylene, and other toxic chemicals;
- Spills and leaks of hydraulic fracking fluid have already contaminated numerous drinking water sources;
- Gas explosions have resulted in forced home evacuations and further drinking water contamination;
- Drilling requires thousands of truck trips that contribute to air and noise pollution; and
- Each fracking well produces up to several million gallons of toxic wastewater that must be disposed of.

Nationwide, residents living near fracked gas wells have filed over one thousand complaints of tainted water, severe illnesses, livestock deaths, and fish kills. Complaints, sometimes involving hundreds of households, have risen at the same time as the number of wells has increased. Fracking would be considered a crime, except that it's been exempted from key provisions of the *Safe Drinking Water Act*, the *Clean Air Act*, and the *Clean Water Act*.

Given all this, it's easy to understand why fracking has become an issue for the U.S. environmental health movement. More than 250 communities have already passed resolutions to stop it, and one hundred environmental health groups have called on President Obama to impose a moratorium on new federal permits.[54] Vermont has already banned fracking, and the battle is on in other states, including New Jersey, Ohio, Pennsylvania, and New York, where well-known ecologist and author Sandra Steingraber is leading a major campaign.

What's remarkable is how quickly environmental health groups and communities have come together on this issue. National and state groups are working closely with local ones to coordinate their efforts. Not only does this reveal widespread opposition to fracking, it also provides visible proof of the U.S. environmental health movement's flexibility and adaptability and the importance of coalition building.

Another example of national, state, and local collaboration is the Beyond Coal movement. Led by local activists coordinated by the Sierra Club, it has prevented the construction of 150 proposed coal-fired power plants, shut down well over 100, and stopped the coal industry dead in its tracks—all in less than a decade. According to Lester Brown, of the Earth Policy Institute, the Beyond Coal movement has imposed "a de facto moratorium on new coal-fired power plants."[55] Another environmentalist went further, saying that stopping new coal plants may be "the most significant achievement of American environmentalists since the passage of the *Clean Air Act* and the *Clean Water Act*."[56]

Alarmed by mountain top removal in Appalachia and President George W. Bush's proposal to construct a huge number of new coal-fired power plants across the entire country, community groups took on the powerful coal industry and are winning. The colossal, but mostly unnoticed, achievements of this movement are the result of the coordinated efforts of local grassroots activists. Their ammunition was irrefutable: air pollution and smog resulting from coal burning contribute to asthma and other respiratory diseases; mercury from coal burning causes neurodevelopmental effects in children and contaminates fish; and the toxics found in coal ash have been linked to organ disease, respiratory illness, neurological damage, developmental problems, and cancer. In fact, people living within one mile of unlined coal ash ponds can have a one in fifty risk of cancer[57]—that's more than two thousand times higher than what the EPA considers acceptable. And on top of this, coal burning releases massive amounts of carbon dioxide into the atmosphere, causing climate disruption. There's not a shred of doubt that coal burning can cause serious effects on human health.

By opposing the coal industry, the Beyond Coal movement has taken the conversation about the U.S. energy future out of Washington, DC, and put it into the kitchens and living rooms of America. On its own, this is a considerable achievement, but this movement has done a lot more. By arguing for renewable energy, such as solar and wind, and energy efficiency and conservation, it's proposing safe, practical solutions. Like the anti-incineration activists of the 1980s, the Beyond Coal movement is not merely against something; it's offering positive solutions, and it has already achieved considerable success. By starting with grassroots organizing and then bringing in the coordinating capacities of the Sierra Club, the Beyond Coal movement has won many victories.

BEYOND TOXICS: URBAN PLANNING AND GREEN BUILDING

Although urban planners and green builders aren't as "in your face" as the activists opposing fracking and coal burning, some are working to benefit environmental health. By improving the design of communities and buildings to enhance health, urban planners and green builders are making a contribution to the cause.

Urban planners' interest in environmental health was rekindled in 1984 when the City of Toronto in Canada held the first "healthy city" workshop, called *Healthy Toronto 2000.*[58] Led by visionary public health physician Trevor Hancock, it led to the creation of the Healthy Cities movement. The basic idea was to reconnect urban planners and public health professionals, and to ensure that community design took account of health, quality of life, and social justice. By 2000, the Healthy Cities movement had spread to more than three thousand communities in more than fifty countries. I was one of the organizers of the original Toronto conference, and I've always found it immensely gratifying that what started out as an idea discussed over lunch grew into a global movement.

In the United States, the Healthy Cities movement was called the Healthy Communities movement, in recognition of the fact that all communities, not just cities, should be healthy. The U.S. movement also had a distinctive American flavor that embraced this country's traditions of local democracy and citizen governance.[59] Flourishing in the 1990s and early 2000s, it's still active today and is now part of urban planning. Raising awareness about the relationship between the built environment and well-being, it's helping to advance environmental health.

Even though the discipline of urban planning doesn't have many formal links with the U.S. environmental health movement, it could become an important ally and partner in the future because several studies have shown that urban planning has direct implications for health. One of the first was a national analysis of physical activity, obesity, and chronic disease which found that people who live in sprawling developments are more likely to have high blood pressure, walk less, and weigh more than those who live in compact developments.[60] This was closely followed by *Urban Sprawl and Public Health*,[61] a book written by Howard Frumkin, Larry Frank, and Richard Jackson. Providing the first comprehensive overview of the links between community design, transportation, architecture, and health, it confirmed that suburban living isn't always healthy. Other studies have yielded similar results. Today, urban planners are designing compact, environmentally friendly neighborhoods intended to enhance community vitality and health.

Green builders have stronger links with the U.S. environmental health movement. This is because in 2000 its leaders collaborated with green builders and established the Healthy Building Network (HBN). Working to decrease the harmful effects of the building industry on environmental health, the HBN provides a valuable bridge between the building industry and environmental health groups. In 2001, the HBN and the Environmental Working Group petitioned the Consumer Product and Safety Commission to ban arsenic-treated wood in playground equipment,[62] and by 2003, major manufacturers had agreed. In 2004, the EPA halted the manufacture and sale of arsenic-treated wood for most residential uses. Current HBN initiatives include the Pharos Project, which provides information on the environmental health impacts of building materials, and the Health Product Declaration, which offers a format to report the contents of building materials and associated health information.

Despite the efforts of the HBN, the green building industry's largest professional association, the U.S. Green Building Council (USGBC), hasn't caught up. For instance, the LEED (Leadership in Energy and Environmental Design) program, its voluntary green building certification system, has been criticized for not doing enough to reduce exposure to toxics in building materials.[63] Now that the U.S. construction industry is beginning to recover from the recession of 2008, it's more important than ever for environmental health groups and green builders to work together to reduce the harmful health effects of building materials.

THE SIGNIFICANCE OF FOUNDATION FUNDING

Whether it's collaborating with green builders or lobbying for new legislation, the U.S. environmental health movement's work has been made possible by the financial support of numerous foundations, such as the Beldon Fund (no longer operational), the John Merck Fund, and the New York Community Trust. The significance of foundation funding in the movement's growth and development is difficult to overestimate.

In 1999, the foundations that support environmental health groups were brought together by Michael Lerner, president of Commonweal, to create the Health and Environmental Funders Network (HEFN). Since then, HEFN has provided a forum for them to share information and work together to maximize the impact of their philanthropy. In the first decade of its existence, HEFN worked with more than 250 grant makers from 125 foundations who invested more than $65 million a year in work on environmental health and justice.[64]

The steady and reliable income stream provided by foundations has enabled the U.S. environmental health movement to build the infrastructure necessary to become an effective social movement. In particular, foundation grants have been used to hire professional staff, including scientists and lawyers; rent, equip, and furnish offices; and launch campaigns. It's clear that the U.S. environmental health movement owes a huge debt of gratitude to foundations. Indeed, it's almost impossible to imagine what the movement would look like today without this ongoing source of support. Much of the money has gone to support the work of national coalitions and state organizations and their work on legislative reform, and these investments have paid benefits in terms of new state and local policies on toxic chemicals, even though success at the federal level has been more elusive. By focusing on passing new legislation, it's been easy for foundations to see the results of their investments. Although success isn't guaranteed, a new law or regulation is a clearly identifiable victory; it's a positive and tangible accomplishment that can be seen by anyone.

So the good news is that many national and state groups have received enough support to enable them to become advocates for legislative reform. The not-so-good news is that many small, local groups struggle to survive because so much of the cash is going to larger groups. This has made some local activists resentful. To add to this, many local groups represent people living in poverty, minorities, and other marginalized populations, so they feel as if they're being discriminated against. This bias was confirmed in a 2012 study of foundation funding which concluded, "From 2007–2009, only 15 percent of environmental grant dollars were classified as benefitting marginalized communities, and only 11 percent were classified as advancing 'social justice' strategies, a proxy for policy advocacy and community organizing that works toward structural change on behalf of those who are the least well off politically, economically and socially."[65] The report went on to state, "we can secure more environmental wins by decreasing reliance on top-down funding strategies and increasing funding for grassroots communities that are directly impacted by environmental harms and have the passion and perseverance to mobilize and demand change." In other words, by supporting small, local groups and their collective action tactics, environmental health funders could leverage greater social change.

Some local activists argue that national and state groups have become overly dependent on foundations. Asserting that foundation-dependent groups spend too much time courting their funders, developing grant proposals, working on projects tailored to fit foundation priorities, and writing reports for them, they claim that these groups aren't sufficiently responsive to locally identified issues. To support this allegation, some point to the fact that many state groups are not membership based and do not actively involve the public in identifying priority issues. Maintaining that state and national groups have traded in their visions of social change for secure jobs and regular paychecks, they claim they've ceded control of the movement to foundation boardrooms and executives. Agreeing with this, one local activist told me, "All too often foundations are calling the tune of the environmental health movement toward policy solutions and ignoring grassroots struggles and movement building." On the other hand, state and national groups that benefit from foundation funding refute this accusation saying that they themselves urged foundations to focus on legislative reform. While this may be true, it leaves many local environmental health groups out in the cold and struggling to get by on a shoestring.

Foundations' desire to support legislative work also means that financial support for other types of activities is very scarce. As well as there being little money for local groups and their struggles on local issues, there's very little money to help environmental health groups respond to public requests for environmental health information. This is unfortunate because, as noted in the section on Communication and Getting the Word Out, people often turn to them for advice. Although it's often possible to get funding to develop outreach materials on legislative topics, it's a lot harder to get grants to respond to requests from the public for information. While it's true that information is widely available on the internet, many people want answers from a real, live human being, and few environmental health groups can afford to provide this service. For many foundations, providing financial support to respond to public inquiries doesn't have the same panache as passing new legislation and policies.

This chapter has described some of the organizations and issues that make up the contemporary U.S. environmental health movement. Although its strength lies in state and local groups, the movement's national organizations, and especially its coalitions, play an essential role too. Indeed, over the past fifteen years or so, the U.S. environmental health movement has transformed itself from a loose assemblage of local groups into a complex and sophisticated network of coalitions, alliances, and partnerships, mostly working to oppose toxic chemicals. But the U.S. environmental health movement isn't only an antitoxics movement; it's also active on nanotechnology, electromagnetic fields, fossil fuels, green building, food (see chapter 6), and many other issues.

One of the U.S. environmental health movement's main achievements in the 2000s has been the success of state groups at passing state and local legislation on toxic chemicals. However, this has come at a price because some local activists have accused state groups of focusing on lobbying and ignoring local groups and their struggles on local issues. Although there is some truth to this, new collaborations to oppose fracking and coal-fired power plants demonstrate that national, state, and local groups are working together more closely than ever before. By combining legislative strategies with coalition building and collective action on local issues, the U.S. environmental health movement is making a real difference. In addition to these overarching and well-established strategies for social change, the movement is also using several unique approaches, including making environmental issues personal, advocating for precaution while making scientific arguments, arguing for environmental justice, and trying to change the economic system. These four strategies are discussed in the following chapters.

NOTES

1. Jessica Schifano, Joel Tickner, and Yve Torrie. *State leadership in formulating and reforming chemicals policy: Actions taken and lessons learned.* Lowell Center for Sustainable Production, University of Massachusetts at Lowell (2009). Available at: http://www.chemicalspolicy.org/downloads/StateLeadership.pdf. Accessed October 31, 2012.

2. Michael Belliveau. The drive for a safer chemicals policy in the United States. *New Solutions: A Journal of Environmental and Occupational Health Policy* 21(3): 359–386 (2011).

3. Michael Belliveau. The drive for a safer chemicals policy in the United States. *New Solutions: A Journal of Environmental and Occupational Health Policy* 21(3): 359–386. p. 374 (2011).

4. Thomas Linzey and Anneke Campbell. *Be the change: How to get what you want in your community.* Layton, UT: Gibbs Smith (2009).

5. Environmental Working Group. *Shopper's guide to pesticides.* Available at: http://www.foodnews.org/http://www.foodnews.org. Accessed October 31, 2012.

6. Collaborative on Health and the Environment. Available at: http://www.healthandenvironment.org. Accessed October 31, 2012.

7. Archive of European Integration. Commission working document SEC (1998) 1986 final. Available at: http://aei.pitt.edu/view/eudocno/SEC_=2898=29_1986_final.html. Accessed October 31, 2012.

8. Chemicals Policy & Science Initiative. *REACH history.* Lowell Center for Sustainable Production. University of Massachusetts at Lowell. Available at: http://www.chemicalspolicy.org/archives.reach.timeline.php. Accessed October 31, 2012.

9. *Copenhagen chemicals charter.* Published by the European Environmental Bureau, the European Consumers' Organisation, the Danish Consumer Council, the Danish Society for the Conservation of Nature, and the Danish Ecological Council (2000). Available at: http://www.chemicalspolicy.org/downloads/Copenhagen%20Charter.pdf. Accessed October 31, 2012.

10. Commission of the European Communities. *White paper: Strategy for a future chemicals policy. COM (2001) 88 Final.* February 27, 2001. Available at: http://eur-lex.europa.eu/LexUriServ/LexUriServ.do?uri=COM:2001:0088:FIN:EN:PDF. Accessed October 31, 2012.

11. World Wide Fund for Nature. *Detox: Campaigning for safer chemicals.* (January 2007). Available at: http://wwf.panda.org/what_we_do/how_we_work/policy/wwf_europe_environment/initiatives/chemicals/detox_campaign. Accessed October 31, 2012.

12. *The Louisville charter for safer chemicals: A platform for creating a safe and healthy environment through innovation.* Available at: http://www.louisvillecharter.org. Accessed October 31, 2012.

13. Ken Geiser, Mark Rossi, and Cathy Crumbley. The Louisville charter: The NGO blueprint for new chemicals policy. *New Solutions: A Journal of Environmental and Occupational Health Policy* 17(3):167–171 (2007).

14. *About safer states.* Available at: http://www.saferstates.com/about/index.html. Accessed October 31, 2012.

15. SAFER States. Available at: http://www.saferstates.com/2012/01/safer-states-2012-legislation.html. Accessed February 12,2013.

16. Safer chemicals, healthy families. Available at: http://www.saferchemicals.org. Accessed October 31, 2012.

17. *Safe Chemicals Act of 2010.* Available at: http://www.govtrack.us/congress/bill.xpd?bill=s111-3209. Accessed October 31, 2012.

18. *Toxic Chemicals Safety Act of 2010.* Available at: http://www.opencongress.org/bill/111-h5820/show. Accessed October 31, 2012.

19. Chloe E. Bird. Gender, household labor, and psychological distress: The impact of the amount and division of housework. *Journal of Health and Social Behavior* 40(1) (1999).

20. Campaign for Safe Cosmetics Survey, June 2004. Available at: http://www.ewg.org/skindeep/2004/06/15/exposures-add-up-survey-results/. Accessed March 1, 2012.

21. Breast Cancer Fund. *State of the evidence: The connection between breast cancer and the environment, sixth edition.* (2010). Available at: http://www.breastcancerfund.org/publications. Accessed October 31, 2012.

22. European Commission. *State of the art assessment of endocrine disrupters*. Available at: http://ec.europa.eu/environment/endocrine/documents/studies_en.htm. Accessed October 31, 2012.

23. Anna M. Soto and Carlos Sonnenschein. Environmental causes of cancer: endocrine disruptors as carcinogens. *National Review of Endocrinology* 6:363–370 (2010).

24. James T. Brophy, Margaret M. Keith, Andrew Watterson, Robert Park, Michael Gilbertson, Eleanor Maticka-Tyndale, Matthias Beck, Hakam Abu-Zahra, Kenneth Schneider, Abraham Reinhartz, Robert DeMatteo, and Isaac Luginaah. Breast cancer risk in relation to occupations with exposure to carcinogens and endocrine disruptors: a Canadian case-control study *Environmental Health* 11:87 (2012).

25. U.S. Department of Labor. *General facts on women and job based health*. Available at: http://www.dol.gov/ebsa/newsroom/fshlth5.html. Accessed October 31, 2012.

26. *BlueGreen Alliance*. Available at: http://www.bluegreenalliance.org. Accessed October 31, 2012.

27. National Religious Partnership for the Environment. *Why is the environment a religious concern?* Available at: http://www.nrpe.org/why/index.html. Accessed October 31, 2012.

28. *Evangelical Environmental Network*. Available at: http://creationcare.org/. Accessed October 31, 2012.

29. Letter to Lisa Jackson, Administrator of the EPA from the Most Reverent Stephen E. Blaire, Chairman, Committee on Domestic Justice and Human Development, U.S. Conference of Catholic Bishops (June 20, 2011). Available at: http://www.usccb.org/about/general-counsel/rulemaking/upload/comments-to-epa-on-mercury-2011-06.pdf. Accessed October 31, 2012.

30. Evangelical Environmental Network. *Fact Sheet: Mercury and the unborn child*. Available at: http://creationcare.org/mercury. Accessed October 31, 2012.

31. National Council of Churches. *Resolution on environmental health*. Washington, DC (2009). Available at: http://www.ncccusa.org/NCCpolicies/environmentalhealth.pdf. Accessed October 31, 2012.

32. *Faith groups ask Congress for safe chemicals policy* (June 29, 2010). Available at: http://www.wfn.org/2010/06/msg00202.html. Accessed October 31, 2012.

33. Coalition on the Environment and Jewish Life. *Global warming: A Jewish response*. (2000). Available at: http://coejl.org/resources/global-warming-a-jewish-response-2000. Accessed October 31, 2012.

34. Evangelical Climate Initiative. *Climate change: An evangelical call to action*. (2006). Available at: http://www.npr.org/documents/2006/feb/evangelical/calltoaction.pdf. Accessed October 31, 2012.

35. Catholic Coalition on Climate Change. Available at: http://www.catholicsandclimatechange.org/. Accessed October 31, 2012.

36. *Interfaith declaration on climate change*. (2009). Available at: http://www.interfaithdeclaration.org. Accessed October 31, 2012.

37. *Statement of the World's Religious Traditions on Climate Change* (2009). Available at: http://www.religionsforpeace.org/initiatives/protect-earth. Accessed March 1, 2012.

38. The Project on Emerging Nanotechnologies. A Project of the Woodrow Wilson International Center for Scholars and the Pew Charitable Trusts. Available at: http://www.nanotechproject.org/inventories/consumer. Accessed October 31, 2012.

39. David Hwang. Nanomaterials state of the market Q1 2009. Cleantech's Dollar Investments, Penny Returns Lux Research. New York (2009).

40. Maureen R. Gwinn and Val Vallyathan. Nanoparticles: Health effects—pros and cons. *Environmental Health Perspectives* 114(12): 1818–1825 (2006).

41. Jessica Marshall. Draft guidelines for nanomedicine unveiled. *Nature*. Published Online, September 28, 2011. Available at: http://www.nature.com/news/2011/110928/full/news.2011.562.html. Accessed October 31, 2012.

42. Principles for the oversight of nanotechnologies and nanomaterials. (2008). Available at: http://www.icta.org/files/2012/04/080112_ICTA_rev1.pdf. Accessed October 31, 2012.

43. Jennifer Sass. *Nanotechnology's invisible threat: Small science, big consequences*. Natural Resources Defense Council Issue Paper. New York. May 2007. Available at: http://www.nrdc.org/health/science/nano/nano.pdf. Accessed October 31, 2012.

44. Environmental Defense Fund-DuPont Nano Partnership. *Nano risk framework*. June 2007. Available at: http://www.edf.org/documents/6496_Nano%20Risk%20Framework.pdf. Accessed October 31, 2012.

45. Nancy Wertheimer and Ed Leeper. Electrical wiring configurations and childhood cancer. *American Journal of Epidemiology* 109(3) : 273–283 (1979).

46. Food and Drug Administration. No evidence linking cell phone use to risk of brain cancer. Available at: http://www.fda.gov/ForConsumers/ConsumerUpdates/ucm212273.htm. Accessed October 31, 2012.

47. IARC Working Group on the Evaluation of Carcinogenic Risks to Humans. *Non-ionizing radiation, Part 1: Static and extremely low-frequency (ELF) electric and magnetic fields*. Lyon, IARC, 2002 (Monographs on the Evaluation of Carcinogenic Risks to Humans, 80).

48. World Health Organization. *Extremely low frequency fields environmental health*. Criteria Monograph No. 238. 2007.

49. *BioInitiative report: A rationale for a biologically-based public exposure standard for electromagnetic fields (ELF and RF)*. Available at: http://www.bioinitiative.org. Accessed October 31, 2012.

50. The Interphone Study Group. Brain tumor risk in relation to mobile telephone use: Results of the INTERFONE case-control study. *International Journal of Epidemiology* 39(3): 675–694 (2010).

51. Nancy Evans, Cindy Sage, Molly Jacobs, and Richard Clapp. *Radiation and cancer: A need for action*. Lowell, MA: Lowell Center for Sustainable Production, University of Massachusetts Lowell (2009). Available at: http://sustainableproduction.org/downloads/RadiationandCancer.pdf. Accessed October 31, 2012.

52. San Francisco passes cell phone radiation law. MSNBC, June 16, 2010. Available at: http://www.msnbc.msn.com/id/37728479/ns/technology_and_science-wireless. Accessed October 31, 2012.

53. Sharon Guynup. *The fracking industry buys Congress*. Environmental News Service, February 16, 2012. Available at: http://ens-newswire.com/2012/02/16/the-fracking-industry-buys-congress/. Accessed October 31, 2012.

54. Letter available at: http://static.ewg.org/pdf/fracking_letter_to_POTUS_03061202.pdf. Accessed October 31, 2012.

55. Lester Brown. *Plan B updates*. November 2, 2011. Available at: http://www.earth-policy.org/plan_b_updates/2011/update101. Accessed October 31, 2012.

56. As quoted in: Mark Hertsgaard. How a grassroots rebellion won the nation's biggest climate victory. *Mother Jones*. April 2, 2012. Available at: http://www.motherjones.com/environment/2012/04/beyond-coal-plant-activism?page=1. Accessed October 31, 2012.

57. RTI. *Human and ecological risk assessment of coal combustion wastes*. Draft Prepared for U.S. EPA Office of Solid Waste (August 2007). Available at: http://earthjustice.org/sites/default/files/library/reports/epa-coal-combustion-waste-risk-assessment.pdf. Accessed October 31, 2012.

58. Trevor Hancock. The evolution, impact, and significance of the healthy cities/communities movement. *Journal of Public Health Policy* Spring 1993: 5–18.

59. Tyler Norris and Mary Pitmann. The healthy communities movement and the coalition for healthier cities and communities. *Public Health Reports* 115:118–124 (2000).

60. Barbara McCann and Reid Ewing. *Measuring the health effects of sprawl*. Smart Growth America, Surface Transportation Policy Project. September 2003. Available at: http://www.smartgrowthamerica.org/report/HealthSprawl8.03.pdf. Accessed October 31, 2012.

61. Howard Frumkin, Lawrence Frank, and Richard Jackson. *Urban sprawl and public health: Designing, planning and building for healthy communities*. Washington DC: Island Press (2004).

62. Healthy Building Network, Environmental Working Group petition Consumer Product Safety Commission to ban sale of arsenic in pressure-treated wood: poisoned playgrounds. Available at: http://www.ewg.org/reports/poisonedplaygrounds. Accessed October 31, 2012.

63. Environment and Human Health, Inc. *LEED certification: Where energy efficiency collides with human health.* (2010). Available at: http://www.ehhi.org/reports/leed/LEED_report_0510.pdf. Accessed October 31, 2012.

64. Michael Lerner, Marni Rosen, Anita Nager, Pete Myers, and Kathy Sessions. Minding the environmental health gap: HEFN marks 10 years of progress. *The Environmental Grantmakers Association Journal* Fall 2009: 9–11.

65. Sarah Hansen. *Cultivating the grassroots: A winning approach for environment and climate funders.* National Committee for Responsive Philanthropy. (2012) p. 1.

Chapter Six

Making Environmental Issues Personal

When I give talks on environmental health, I often start by asking the audience if I can do a quick survey with them. Of course, they always agree. I begin by saying, "How many of you know someone living with cancer?" Usually, about half the people in the room raise their hands. Then I ask, "How many of you know someone living with heart disease?" Again, about half raise their hands. Finally, I ask, "How many of you know someone living with asthma?" By this time, almost everyone has raised their hands and people are looking around the room, expressing surprise at the widespread nature of these three health conditions. Next I say, "Did you know that according to the World Health Organization about 20 percent of all cancers, roughly 15 percent of all heart disease, and almost half of all asthma are caused by environmental pollution?"[1] At this point, surprise turns into concern as the real-life health consequences of environmental pollution hit home.

This "survey" personalizes environmental issues. By encouraging people to reveal the environmentally related diseases and disabilities they're coping with, it makes the environment deeply meaningful and something that everyone can relate to. This is an extremely powerful experience that always gives rise to feelings of empathy and compassion, as well as demands to stop pollution.

Making environmental issues personal is the environmental health movement's defining strategy, and it's the subject of this chapter. Over the years, the movement has found many ways to put a human face on environmental issues. By gaining support from the victims of environmentally related disease and their caregivers; drawing attention to children's environmental health; and highlighting the presence of toxic chemicals in food, consumer products, and our own bodies, it's made environmental issues very real and immediate to the public.

This strategy took advantage of the young environmental movement's decision to ignore the urban environmental health concerns of many Americans and leveraged them into creating a new social movement. But it wasn't only that environmentalists disregarded urban environmental health issues, it was also that when they did talk about health—which wasn't very often—they used the cold, hard language of science. For instance, when environmentalists linked environmental pollution with cancer and genetic mutation, they used quantitative risk assessment information. This abstract data may have been sufficient to convince Congress to pass the environmental legislation of the 1970s, but it didn't speak to the American public in ways that they understood, and it certainly didn't speak to people's hearts and souls. By failing to translate its issues into language that would resonate with ordinary people, the environmental movement missed a valuable opportunity. But in 1978 Lois Gibbs seized it with both hands (see chapter 4). Talking about her children's environmental health problems on national TV, she transformed public debate about the environment and gave birth to the environmental health movement.

Most national environmental groups now recognize the need to make health-based arguments, and many do it by personalizing their issues. Using the strategy originally developed by local environmental health groups, they are now attracting supporters and strengthening their case. As a consequence, the demarcation between environmental groups and environmental health groups is no longer clear.

GAINING SUPPORT FROM PEOPLE AFFECTED BY ENVIRONMENTALLY RELATED DISEASES

The U.S. environmental health movement makes environmental issues personal in several ways, but perhaps the most important is by gaining support from people affected by environmentally related diseases and disabilities from across the entire country. Simply drawing attention to local pollution problems, however egregious, is not enough. Movement leaders know that the American public needs to understand that environmental health is a national problem; it's not confined to specific communities, such as Love Canal and Times Beach; everyone is at risk, no matter where they live. Whether it's listening to a cancer survivor from Alaska talk about how she has coped with the disease, or a mom from Florida talk about her child's learning disabilities, movement leaders know that hearing the stories of real people who are dealing with real illnesses is very powerful. Putting a human face on environmental problems, this strategy is more compelling than listening to a scientist or a lobbyist talk about the results of a theoretical risk assessment. It's not even in the same league. This national approach has made environmental health into an issue that transcends zip codes and geography. Today, no one can say "pollution doesn't affect me" or "I'm not at risk." Now it's obvious that everyone is potentially affected.

Representatives of those suffering from diseases and disabilities can be powerful social change agents. Just think of what HIV/AIDS activists achieved in the 1980s and 1990s. They knew how to mobilize public concern and win political battles. By making HIV/AIDS into a personal issue, they overcame the social stigma attached to this disease and secured millions of dollars for research that led to effective treatments.

Although HIV/AIDS activists are probably the most successful health advocates in history, the representatives of people suffering from environmentally related illnesses, and their organizations, are extremely capable too. Recognizing this, the U.S. environmental health movement has reached out to many health-affected populations and the groups that represent them, including national organizations representing people with learning and developmental disabilities, breast cancer, multiple chemical sensitivity—a chronic medical condition characterized by symptoms attributed to low-level chemical exposure—and asthma.

Several national organizations working on learning and developmental disabilities have been very active in environmental health since the mid-2000s, including the American Association on Intellectual and Developmental Disabilities (AAIDD), the Autism Society of America, and the Learning Disabilities Association (LDA). Largely as a result of the outreach efforts of the Collaborative on Health and the Environment (CHE), these organizations are now partners in the U.S. environmental health movement. And there's a good reason for this. Exposure to numerous toxic chemicals, including lead, mercury, and PCBs, has been associated with neurodevelopmental deficits. Mostly affecting children, the burden of learning and developmental disabilities is huge. According to one analysis of National Health Interview Survey data, the prevalence of developmental disabilities, such as autism spectrum disorder and ADD/ADHD, rose from 12.84 percent to 15.04 percent between 1997 and 2008, and between 2006 and 2008 developmental disabilities were reported in about one in six children in the United States.[2] What makes learning and developmental disabilities even more tragic is that many, especially those caused by environmental contaminants, can be prevented.

Many people affected by learning and developmental disabilities, as well as the organizations that represent them, are part of the Collaborative on Health and Environment's Learning and Developmental Disabilities Initiative (LDDI), which lists well over five hundred participants on its web site.[3] In 2008, the LDDI released a Scientific Consensus Statement and a Policy Statement on the environmental agents associated with neurodevelopmental disorders. Signed by many nationally recognized scientists, health professionals, academics, environmental health and justice groups, and members of the public, these two statements advocate chemical policy reform, public education, and research to reduce exposures to substances implicated in neurodevelopmental disorders.[4]

Breast cancer groups have been supporters of the U.S. environmental health movement for even longer than learning and developmental disability groups. Starting in the early 1990s, several breast cancer groups emphasized the need for preventing this disease by reducing exposure to toxic chemicals. Breast cancer organizations in three different regions of the country have been particularly active: Boston/Cape Cod; Long Island, New York; and the San Francisco Bay area. These regions all have higher incidences of breast cancer than the national average and are home to highly educated, politically active populations of women. Notable groups include the Massachusetts Breast Cancer Coalition and the Silent Spring Institute in Boston/Cape Cod, the Long Island Breast Cancer Network (although it no longer exists), and the Breast Cancer Fund and Breast Cancer Action in the San Francisco Bay area. While these groups have their own unique identities, the Breast Cancer Fund stands out because of its scientific reports. Over the years, the Fund has published many editions of a report reviewing the latest scientific evidence that environmental agents cause breast cancer. Called *State of the Evidence: The Connection between Breast Cancer and the Environment*, this report has been invaluable to the environmental health movement and women's groups across the country and internationally.

The high level of concern about breast cancer and toxic chemical exposures isn't surprising because there's mounting evidence that environmental contaminants play a role in this life-threatening disease and rates are rising rapidly. Between 1973 and 1998, breast cancer incidence rates in the United States increased by more than 40 percent. Today, a woman's lifetime risk of breast cancer is about one in eight,[5] several studies have shown a link between environmental endocrine disruptors and breast cancer risk,[6, 7] and at least one has shown an increase in breast cancer rates associated with increased exposure to toxic chemicals.[8] A staggering 216 chemicals and radiation sources have been listed by international and national regulatory agencies as being experimentally implicated in breast cancer causation.[9] Confirming what the Breast Cancer Fund and other environmental health groups have been saying for years, in February 2013 a federal committee of agency representatives and scientists conculded that environmental factors, including toxic chemicals, increase the risk of breast cancer and called for a national prevention strategy.[10]

A third health-affected population active in the U.S. environmental health movement is people suffering from multiple chemical sensitivities (MCS). The number of people living with MCS is large but unknown because it can be difficult to define and diagnose this set of health problems. One of the first organizations for people with MCS was the National Center for Environmental Health Strategies (NCEHS). Founded in 1984, it advocated for people with MCS. Although the NCEHS no longer exists, organizations such as the Multiple Chemical Survivors and the Chemical Sensitivity Disorders Association carry on its work. In the 1990s, new populations, including veterans and women with silicone breast implants, started to report MCS and added to concerns about the effects of low exposures to toxics. In particular, veterans of the 1991 Gulf War have reported a wide range of acute and chronic symptoms, including fatigue, musculoskeletal pain, cognitive problems, skin rashes, and diarrhea.

Although not directly related to MCS, some veterans, service members, and their families are concerned about toxic exposures on military bases. Given the history of contamination and environmental health problems at many facilities, this concern is quite justified. For instance, residents of the Marine Corps base at Camp Lejeune, North Carolina, were exposed to contaminated drinking water for decades and suffered from numerous health problems including various forms of cancer. After many studies and investigations, the federal government took responsibility and agreed that at least three former service members' cancers were caused by toxic exposures. Then in 2012 Congress passed the *Janey Ensminger Act* in honor of Janey Ensminger, who lived at the base and died of cancer at age nine. This *Act* authorizes medical care for military and family members who lived at Camp Lejeune between 1957 and 1987 and developed health problems linked to the water contamination. Although tragically sad, the problems at Camp Lejeune and other military facilities, combined with evidence of MCS in veterans, are raising awareness about environmental health in military, and former military, personnel. And this is helping to broaden the environmental health movement's base of support to include people who had not been participants before.

A fourth health-affected population is people suffering from asthma and other respiratory problems linked with air pollution. At a national level, the American Lung Association (ALA) has expressed its concern about air quality because, as noted at the beginning of this chapter, almost half of all asthma is caused by air pollution. Compounding this, the lifetime prevalence of asthma in the United States is very high; in 2011, it was 12.9 percent.[11] To try to prevent this heavy toll on health, the ALA and its state affiliates are working to improve U.S. air quality. At the national level, the ALA has programs to educate the public on air quality at home, in schools, at work, and outdoors. It also advocates for stronger federal regulations on air pollution and publishes an annual *State of the Air* report. At the state level, programs such as the American Lung Association in Washington's Master Home Environmentalist program offers free home assessments.

Support from organizations that represent people affected by environmentally related diseases has helped the U.S. environmental health movement enormously, but it can be damaging when health groups are unsympathetic or even hostile. Take the difference between the American Cancer Society (ACS) and its Canadian counterpart, the Canadian Cancer Society (CCS). The ACS has consistently downplayed the role of environmental factors in cancer, while supporting corporations that manufacture pharmaceuticals, pesticides, and other chemicals. Fixated on diagnosis and treatment, the ACS has repeatedly failed to support policies to reduce exposure to carcinogenic chemicals.[12, 13, 14] In contrast, the CCS is a staunch supporter of environmental health. A statement on its web site reads: "We believe that you shouldn't be exposed to substances that cause cancer at work, at home or in your environment. . . . Substances that cause cancer should be replaced with safer alternatives."[15] The CCS has also taken strong positions in support of the right to know and reducing occupational exposures to carcinogens, and opposing the use of lawn and garden pesticides and asbestos. Its work has done a lot to raise awareness about environmental health and enhance the standing of Canadian activists. The failure of the ACS to support the U.S. environmental health movement is damaging because it ignores the scientific facts and weakens the movement's credibility.

WORKING WITH CAREGIVERS—NURSES

As well as reaching out to people suffering from environmentally related diseases and the organizations that represent them, the U.S. environmental health movement has intentionally courted caregivers, especially nurses and physicians. Indeed, for the past fifteen years, nurses have been in the forefront of efforts to improve environmental health. But nurses' concerns have a much longer history.

Florence Nightingale (1820–1910) is often credited with introducing environmental health into the nursing profession. Her book *Notes on Nursing*,[16] first published in 1860, describes how air quality, water quality, noise, light, and nutrition affect health and well-being. Its discussion of air quality is among the earliest modern thinking about the relationship between the environment and health. Commenting that coal used for home heating contributed to poor air quality, the book drew attention to indoor as well as outdoor pollution. Indeed, Nightingale was so adamant about the importance of indoor air quality and ventilation that she prioritized it in *Notes on Nursing* as the "first cannon of nursing."[17]

As a result of Florence Nightingale's work, preventing environmentally related disease became part of the nursing profession. In addition to Florence Nightingale herself, other nurses who were advocates for environmental health include Lillian Wald (1867–1940), Josephine Baker (1873–1945), Frances Perkins (1880–1965), and Harriet Hardy (1905–1993). But over time, its role declined as nursing became increasingly specialized and treatment oriented. Fortunately, in the mid-1990s this trend was reversed and nurses became leaders in environmental health once again.

This resurgence was prompted by a landmark report called *Nursing, Health and the Environment*, published in 1995 by the Institute of Medicine.[18] The report described the importance of environmental health protection in nursing, saying: "Environmental health hazards, including those in the work environment, are ubiquitous, often insidious, and generally poorly understood. As such, they are of . . . fundamental importance to health care providers." It went on to point out nurses' lack of training, saying: "They are also the largest group of professional health care providers in the United States: an estimated 2.2 million. . . . Yet the vast majority of nurses have had no formal training in occupational or environmental health."

Some nursing schools now include environmental health as part of the curriculum and several continuing education programs offer courses on the subject. But nurses have done a lot more than just educate themselves; they have also become strong advocates. The American Nurses Association (ANA) passed its first resolution on environmental health in 1997,[19] and since then it's passed many more. In fact, no other professional health association has passed more resolutions, except for the American Public Health Association. In 2008, fifty nursing leaders came together to create a new organization, the Alliance of Nurses for Healthy Environments.[20] The Alliance provides an online networking forum for nurses interested in environmental health, covering research, practice, education, and policy/advocacy and has helped to raise the profile of environmental health within the profession.

Nurses are especially important partners in the U.S. environmental health movement for several reasons. Perhaps most importantly, they have direct contact with patients, families, and communities. Indeed, nurses are often the first and only point of contact for people seeking medical care. With access to homes, schools, and workplaces, they are in an ideal position to influence people's behavior and help prevent environmentally related diseases. Moreover, because they work with the most vulnerable populations—babies and young children, the frail and elderly, and low-income and minority groups—nurses can help to reduce the health disparities associated with environmental exposures. People usually listen to what nurses say because they are trusted and respected members of society. Not only does this mean that ordinary people often do what they say, it also means that they make excellent public policy advocates. When nurses speak, legislators listen; so they are in an ideal position to make the case for regulations on environmental health.

WORKING WITH CAREGIVERS—PHYSICIANS

From an institutional perspective, the medical profession has been much less supportive of the U.S. environmental health movement than the nursing profession, but like nurses, physicians have a long-standing concern about environmental health. In fact, their interest far predates nurses'. Ever since Hippocrates argued that human health is affected by environmental quality, physicians have played a role in protecting environmental health. Just think of Percivall Pott, the first person to identify an environmental carcinogen (chimney soot); Rudolf Virchow, who advocated for legislation to protect environmental public health; and John Snow, who had the handle of the Broad Street water pump removed to reduce the spread of cholera in London (see chapter 1).

But despite the pioneering work of these and other physicians, the medical profession has not wholeheartedly supported the U.S. environmental health movement. One reason for this may be its views on how to improve health. With a belief in curative approaches, the medical profession devotes its energies to treating people after they've gotten sick. In fact, it could be said that the United States has a system for sickness care, not health care. This after-the-fact approach ignores the importance of preventing diseases and disabilities. And because of this, it also overlooks the fact that improvements in environmental quality would reduce the rates of many environmentally related diseases. Even today, many physicians are skeptical about the influence of the environment on health and well-being.

Notwithstanding its focus on curative approaches and after-the-fact treatment, the medical profession has played a role in environmental health. In 1963, the year after the publication of *Silent Spring*, the American Medical Association (AMA) established a Committee on Environmental Health, and in 1964, it held its first national congress on the subject, which included papers on air pollution and disease, the role of physicians in "community control activities," pesticide poisoning, radioactive contamination, and water pollution control.[21] Since then, the AMA has passed resolutions on the management and disposal of radioactive and chemical wastes, the health effects of global warming, the reform of the *Toxic Substances Control Act* (1976), the use of ecologically sustainable and nontoxic products in health-care facilities, lead, and other environmental health issues.

Several professional medical associations and institutions have recommended that environmental health should play a larger role in medical education. For instance, the Institute of Medicine has proposed including environmental health in all levels of medical training,[22] and the American Medical Association has encouraged physician educators to devote more attention to it and encouraged physicians to educate themselves about pesticide-related illnesses.[23, 24] As well, several medical associations have endorsed a National Environmental Education Foundation Program position statement calling for the further integration of environmental health into medical and other health-care provider education.[25]

But despite this, very little has actually been achieved. It's true that the 2007 accreditation requirements for residency training programs in pediatrics and family medicine include expectations for learning about environmental illness and injury, that a handful of medical curricula include environmental health, and that there are a few occupational and environmental health residency programs.[26] There are also training programs run by nonprofit organizations, such as Physicians for Social Responsibility. But this isn't much, especially when one considers the influence of the medical profession in U.S. society, the role of physicians in health-care decisions, and the fact that there are more than 650,000 of them in the country.

Several professional medical associations have gone further. Mostly related to reproductive and children's health, they include the American Academy of Pediatrics (AAP), the Ambulatory Pediatric Association (APA)—now called the Academic Pediatric Association—the American Congress of Obstetricians and Gynecologists (ACOG), the American Society for Reproductive Medicine (ASRM), the Association of Reproductive Health Professionals (ARHP), and the American College of Preventive Medicine (ACPM). The American Academy of Pediatrics has a very active Council on Environmental Health which lobbies for policies to protect children's environmental health, and in 1999, it published a *Handbook of Pediatric Environmental Health*, which is now in its third edition.[27] As well, pediatricians led the way in establishing a network of Pediatric Environmental Health Specialty Units (PEHSUs) in hospitals and universities. Located in the United States, Canada, and Mexico, PEHSUs offer advice on prevention, diagnosis, management, and treatment of environmental health effects in children. Related to the PEHSUs, in 2001 the Academic Pediatric Association established a National Fellowship Program in Pediatric Environmental Health and proposed competencies for pediatric environmental health specialists.[28] Although not quite as engaged, the American College of Preventive Medicine has an Environmental Health Committee, participates in the Children's Environmental Health Network, and has called for the overhaul of the *Toxic Substances Control Act* (1976).[29]

Meanwhile, a few regional medical associations are supportive of environmental health. The San Francisco Medical Society has been active on the subject for many years, working to educate physicians and advocating for new legislation. And although it doesn't highlight environmental health, the New York Academy of Medicine does call attention to urban health. Its tagline reads "At the heart of urban health since 1847." It hosted a major environmental health conference in 2012.

But the active support of these organizations is not enough, and the reluctance of the entire medical profession to get behind the U.S. environmental health movement is distressing. Like nurses, physicians are trusted members of society, and they could do a lot to help prevent environmentally related diseases and disabilities and lobby for stronger legislation.

In the absence of strong support, Physicians for Social Responsibility (PSR) has stepped into the breach. Comprised of about thirty-two thousand medical and health professionals and citizen members, PSR is one of the most prominent nongovernmental groups working on environmental health today. It was originally founded in 1961, and played a key role in the campaign to end atmospheric nuclear weapons testing. Over time, PSR's efforts have grown into an international movement, and in 1980, it helped to establish the International Physicians for the Prevention of Nuclear War (IPPNW). In 1985, PSR was jointly awarded the Nobel Peace Prize with IPPNW. In the early 1990s, PSR broadened its scope to include other issues, and in 1993 its Greater Boston Chapter collaborated with MIT, the Harvard School of Public Health, and Brown University to publish *Critical Condition*,[30] one of the first books on global environmental health. The Greater Boston Chapter of PSR has been especially active, releasing a report on children's environmental health[31] and providing trainings on the subject, as well as raising public awareness about aging and environmental health.[32] Even though the medical profession has been reluctant to wholeheartedly support the U.S. environmental health movement, PSR does.

PROTECTING CHILDREN'S ENVIRONMENTAL HEALTH

One of the most important ways in which the U.S. environmental health movement has personalized the effects of toxic chemicals is by highlighting the risks to the most vulnerable group in society—children. This is an extremely powerful issue because people care very deeply about children and their health. Almost everyone is a parent, a grandparent, an aunt or uncle, a teacher or a caregiver, so the idea that children's health can be harmed by toxic substances is extremely troubling. Cutting across race, gender, religion, socioeconomic and regional differences, children's environmental health has attracted many people and organizations to the movement.

But this issue goes far beyond people's feelings and emotions; there's actually an enormous body of scientific research showing that children face significantly greater environmental health risks than adults. Children aren't simply little adults when it comes to toxic chemicals and other environmental exposures. They eat more, drink more, and breathe more in proportion to their body weight than adults, so pound-for-pound their exposures are much greater. As well, children's behaviors can expose them to more chemicals and microorganisms than adults. For instance, babies and young children often play on the ground and are exposed to contaminants in soil, carpets, or paved areas. And, as any parent or caregiver knows, they often put things in their mouths. As well, because children are shorter than adults, they can be exposed to larger amounts of ground level air pollutants. On top of all this, children's metabolic and physiological systems are still being formed, so they are especially vulnerable to the developmental effects caused by many toxic substances. Their cells are still multiplying very rapidly, and their nervous, respiratory, reproductive, and immune systems are not fully developed until well after birth. Put it all together, and children face significantly greater risks and are much more vulnerable than adults. Too young to speak for themselves, they need the environmental health movement to advocate on their behalf.

The U.S. environmental health movement's work on children's environmental health began in 1989, when new mother and Oakland, California, resident, Joy Carlson became concerned about scientific studies showing that children face greater health risks from pesticides in their food (see chapter 4) and about health professionals' lack of knowledge. So together with a few colleagues, she established the Kids in the Environment Project to train California's health professionals on children's environmental health. By 1992, the Project had evolved into the Children's Environmental Health Network (CEHN), with Carlson as its first executive director. In the same year, the Children's Health and Environmental Coalition (CHEC) was formed (now called Healthy Child, Healthy World), and in 1999 these two national organizations were joined by the Institute for Children's Environmental Health and the Partnership for Children's Health and the Environment. All of these organizations played a major role in raising public awareness about children's environmental health and making this subject central to the work of the environmental health movement for more than twenty years.

In 1993, CEHN cosponsored the first national research workshop on children's environmental health, which identified key research needs on the subject. In the same year, the National Research Council's (NRC) study on *Pesticides in the Diets of Infants and Children*[33] confirmed the conclusion of the 1989 Natural Resources Defense Council report *Intolerable Risk: Pesticides in Our Children's Food*[34] (see chapter 4) that regulations on pesticide residues in food did not adequately protect infants and children. Then in 1994, CEHN sponsored the first national symposium on children's environmental health, which created a basic policy and research framework for children's environmental health and galvanized interest at the national level.[35]

Within a few years, the efforts of CEHN, NRDC, and other groups began to pay off when Congress passed the *Food Quality Protection Act* (1996). Although it provided an extra safety margin in the standards for pesticide residues in food to protect children's health, it was rather a hollow victory because it was never really implemented and repealed the Delaney Clause (see chapter 3). Also in 1996, the EPA released a major report on children's environmental health, as well as an ambitious National Agenda.[36] A few months later, in 1997, President Clinton signed an executive order[37] requiring all federal agencies to consider children's health and safety in policy decisions. Although there aren't many legal teeth in it, the executive order provides a tool to educate the public and raise awareness about children's environmental health. Soon after, the EPA created the Office of Children's Health Protection to help implement the National Agenda and the Executive Order. Then in 2000, the *Children's Health Act* authorized the federal government to conduct a major study on the effects of the environment on the health of U.S. children; however, the design of the main study is still being debated some thirteen years after the legislation was passed.

Although these federal initiatives are significant, much remains to be done. In recent years, the U.S. environmental health movement has begun to expand its efforts on children's environmental health by exploring cross-generational environmental health, including the environmental threats to healthy aging.[38] By linking children's environmental health with environmental health at other ages and stages of life, it's drawing attention to the lifecycle impacts of toxic chemicals.

FOOD, GLORIOUS FOOD

Food is another issue that personalizes environmental issues. Grown in the air, water, and soil of the environment, our food becomes part of who we are. It represents nothing less than the transformation of the environment into the cells and tissues of our bodies. For this reason, food is a very intimate and personal matter. But at the same time, it's also a very public matter. We all have to eat, and we often do it with other people. Indeed, sharing a meal is one of the most common of all human rituals. Given that food is both a very private and a very public matter, it's an ideal issue for the environmental health movement.

Author Frances Moore Lappé was one of the first people to talk about food as an environmental issue. By drawing attention to the links between people's food choices and global industrial agriculture, she argued that a vegetarian diet was the best way to protect the environment and reduce world starvation. Her 1971 book *Diet for a Small Planet*[39] called for sustainable agriculture and coined the term "food system." However, beyond convincing some people to stop eating meat, Lappé's ideas fell on deaf ears, and industrial agriculture became an increasingly powerful and profit-hungry sector of the economy. Willing to cut corners and do almost anything to make a dollar, it became thoughtless and irresponsible, as evidenced by issues such as "mad cow" disease and its human counterpart Creutzfeldt–Jakob disease, the spread of genetically modified crops (GMOs), and the use of "pink slime"—a purée of slaughterhouse beef scraps treated with ammonia—to extend hamburger meat.

These and other issues have contributed to a loss of confidence in industrial agriculture and to the emergence of the U.S. food movement. Encompassing many different issues, including organic food, pesticide use, local farmers' markets, "slow food," food justice, GMOs, antibiotic and hormone use in farm animals, the application of toxic sewage sludge to food crops, food packaging and labeling, and aquaculture, the U.S. food movement is actually more of a loose network of local and regional groups and individuals than a coherent social movement. What's remarkable is how quickly these groups have become a significant force in American society. In 2007, there were about 12,500 community-supported agriculture programs,[40] up from just two in the mid-1980s.[41] Similarly, in 1994, there were only 1,755 farmers' markets, but by 2012 there were 7,864.[42] And in 1990, sales of organic foods and beverages in the United States were about $1 billion, and in 2010 they hit nearly $27 billion.[43] Today, big-box stores like Wal-mart and Target sell organic food; ecologically conscious chefs like Alice Waters of Chez Panisse in Berkeley, California, have become superstars; and local food has become a must-have for trendy restaurants across the country.

Arguably, no one has done more to launch the U.S. food movement than author Michael Pollan. His books—*The Omnivore's Dilemma*,[44] *In Defense of Food*,[45] *Food Rules*,[46] and *Cooked*[47]—are a wake-up call for anyone who cares about what they eat. Asking fundamental questions about the ecological and ethical consequences of our dietary choices, Michael Pollan challenges his readers to think more deeply about their food. Other writers such as Eric Schlosser—*Fast Food Nation*[48]—Marion Nestle—*Food Politics*[49] and Raj Patel—*Starved and Stuffed*[50]—have also contributed to the growing food movement. And, of course, First Lady Michelle Obama is a strong supporter. Her 2012 book *American Grown*,[51] sings the praises of vegetable gardens and healthy food.

The rapid growth of the U.S. food movement presents an amazing opportunity for the environmental health movement, and many state and local groups have seized it and are now working with "foodies" to promote local, organic food and encourage urban agriculture. In many poor and minority communities, food justice has become an important issue and groups are working to increase the availability of affordable, fresh food in "food deserts." By planting fruits and vegetables in abandoned lots, schools, hospitals, and prisons, they are raising awareness about nutrition and building healthy, vibrant communities. Meanwhile, national organizations, like the Institute for Agriculture and Trade Policy and the Center for Food Safety, have developed impressive programs to raise public awareness and lobby for legislation.

It's now clear that many Americans are concerned about where their food comes from, as well as how it's produced and processed. It's equally clear that this concern is leading to the creation of alternative, local food economies. But what's less obvious is whether the U.S. food movement can become politically effective. So far, it's focused on building alternatives to the industrial agricultural system, rather than opposing it head-on. This is the easy work. The much harder work that lies ahead is to confront the powerful social, political, and economic interests that dominate the U.S. food system. Unless the food movement is willing to take on the corporations and institutions that control industrial agriculture, it will remain just a well-intentioned, feel-good movement. But if it takes the leap and goes from growing the perfect heirloom tomato to tackling the United States' powerful agricultural interests, it will need to build its political savvy and develop new skills to organize collective action. One of the ways that it could do this is by working more closely with the U.S. environmental health movement.

OPPOSING TOXICS IN CONSUMER PRODUCTS

Although food is an important consumer product in the struggle for environmental health, manufactured products should not be overlooked. Even though some people do not think of consumer products as part of the environment, they are always part of the built environment where most of us spend the vast majority of our time—in homes, cars, schools, and public places. Thanks to the creation of a consumer society, we live in places filled with manufactured products—computers and cell phones, furniture and fabrics, TVs and toasters, washing machines and weed whackers. The list is endless. There's no escaping the fact that we all live with thousands of consumer products.

Like our food choices, decisions about the products we buy are often very personal. Revealing who we are, the things we choose to buy say something about our unique identities and how we'd like to be seen in the world. But many consumer products contain toxic chemicals. Found in children's toys, household cleaners, furniture, carpeting, electronics, food, kitchenware, clothing, shoes, pesticides, paints, cosmetics, pharmaceuticals, and many other items, they pose significant risks to environmental health. Some of the most common toxics include: formaldehyde in building materials, carpeting, and furniture; flame retardants in furniture and electronics; phthalates in plastics; and bisphenol A in metal cans and plastics. Reducing or eliminating them from consumer products would significantly decrease the environmental health risks faced by the American public.

The U.S. environmental health movement has lobbied for the removal of toxics from consumer products for many years. Early examples include the campaigns against lead in paint and gasoline (see chapter 4). Today, national and state groups like the Center for Health, Environment and Justice (CHEJ) and the Washington Toxics Coalition continue to oppose toxics in consumer products and educate the public about how to reduce their exposures. For instance, CHEJ and its partners have developed a searchable database on toxics in over five thousand consumer products called HealthyStuff.org, and the Green Guide web site, now a project of National Geographic, offers about twenty-five guides or videos to help consumers make informed choices.

Many environmental health groups have successfully lobbied for stronger legislation. By making the case for policies on specific types of products and for bans or limits on individual substances in products, they have secured passage of numerous state laws and local ordinances. A study of chemicals policy from 1990 to 2009[52] found that:

- Nineteen states have legislation to prohibit the sale or distribution of packaging containing intentionally added cadmium, lead, mercury, and hexavalent chromium, and to set limits on the concentrations of these substances;
- Three states have executive orders that require state agencies to purchase and use environmentally preferable cleaning products. Several counties and cities have similar policies;
- Ten states, two counties, and two cities have legislation that regulates toxics in children's products or toys;
- Thirty-two states, four counties, and twenty-one cities have enacted or proposed legislation to ban, restrict, or discourage the use of mercury;
- Twelve states have legislation restricting the use of PBDEs in a wide array of products, including building materials, electronics, furnishings, plastics, polyurethane foams, and textiles; and
- Fourteen states, one county, and one city have legislation prohibiting the use of lead in some consumer products, including pipes, wheel weights, fishing tackle, tableware and housewares, cosmetics, jewelry, children's toys and products, candy, lunch boxes, and other novelty consumer items.

But despite these victories at the state and local levels, the federal government has done very little. Most chemicals in products are exempt from the requirements of the *Toxic Substances Control Act* (TSCA). Moreover, the *Consumer Safety Products Act* (CSPA) is weak, and there is no approvals process for chemicals in products, comparable to the process required for drugs. Clearly, there is an urgent need for federal action.

One of the difficulties is that there is hardly any information available on the types and amounts of chemicals in consumer products. Even though the Toxics Release Inventory requires some corporations to report what they release into the environment, there aren't any requirements for them to report what toxics they use in products. This is a gaping hole in the regulatory system. Despite the fact that chemicals are ubiquitous in consumer products, we have no idea of the quantities or types involved. What is known is that the numbers are staggering and dwarf the amounts released to the environment. A 2004 report published by the National Environmental Trust[53] found that for every pound of neurotoxins, carcinogens, and reproductive or developmental toxics that corporations said they released to the environment, they shipped forty-two pounds of the same chemicals as, or in, consumer products.

But although little is known about the types and amounts of toxic substances in consumer products, there's no doubt that we are exposed to toxics in products every day of our lives. Evidence for this comes from a relatively new technology, called XRF or X-Ray Fluorescence. Hand held XRF analyzers can detect the presence of toxic chemicals, such as lead, mercury, brominated flame retardants, and polyvinyl chloride (PVC), in consumer products. Using XRF analyzers, environmental health groups are showing consumers exactly what's in the products they buy and use every day. For instance, the product testing for HealthyStuff.org is done using an XRF analyzer. Coming as a revelation to many, this technology provides a personalized analysis of what people are actually exposed to in their homes and communities.

The work of the U.S. environmental health movement has not only strengthened state and local legislation on toxics in consumer products, it's also helped to change consumer behavior and attracted many new supporters to the cause. In fact, the market for "green" products is exploding. According to the LOHAS (Lifestyles of Health and Sustainability) trade association, approximately 13 to 19 percent of the adults in this country are "LOHAS consumers," and in 2008 they spent an estimated $290 billion on "green" goods and services.[54] Representing significant growth since the 2005 estimates, which estimated spending at $219 billion, these numbers suggest that many are changing their purchasing habits and supporting environmental health.

AND IN PERSONAL CARE PRODUCTS

The most intimate consumer products we buy are the personal care products that we use on our bodies—soaps, washes, shampoos, conditioners, styling gels, mousses, hair dyes, skin creams, lotions, cosmetics, toothpastes, mouthwashes, deodorants, and sunscreens—and many others. By working to remove toxics from personal care products, the U.S. environmental health movement is further personalizing the toxics issue.

Personal care products contain a huge variety of hazardous substances—everything from formaldehyde and butylated hydroxytoluene to parabens and phthalates. About 10,500 ingredients are used in personal care products, and very few have been adequately tested for their safety. In other words, the products we use to take care of ourselves and make us look healthy and attractive actually contain thousands of untested and potentially hazardous substances. Even worse, most people use many personal care products every day. According to one industry estimate, people can use as many as twenty-five different cosmetic and toiletry products on any given day. If each of these products contains ten different ingredients—which isn't unusual—the average person could be directly exposed to more than two hundred different chemical compounds every day of their lives.[55] Furthermore, more than 20 percent of personal care products contain chemicals linked to cancer, 80 percent contain ingredients that commonly contain hazardous impurities, and 56 percent contain penetration enhancers that help deliver ingredients deeper into the skin.[56]

Given these facts, it's not surprising that the U.S. environmental health movement has had an active campaign opposing the use of chemicals in personal care products for many years. Called the Campaign for Safe Cosmetics, it targets a very profitable industry worth about $280 billion a year worldwide.[57] This coalition of women's, public health, labor, environmental health, justice, and consumer rights organizations advocates the elimination of substances linked to cancer, birth defects, and other health problems from personal care products.

The Campaign for Safe Cosmetics has lobbied hard for legislation such as the federal *Safe Cosmetics Act*, which was introduced into Congress in 2011, and the *California Safe Cosmetics Act* (2005), which requires cosmetics companies to publicly report their use of chemicals linked to cancer and birth defects. To inform consumers about which toxic chemicals are present in which personal care products, the Campaign also developed *Skin Deep*, the world's largest publicly accessible online database of chemicals in personal care products, in collaboration with the Environmental Working Group. This database contains safety ratings for nearly a quarter of all cosmetics and toiletries on the market—over sixty-nine thousand products containing more than eighty-five hundred ingredients. It is the largest and most popular product safety database in the world. Since its launch in 2004, it has been searched more than 260 million times.

Building on the success of *Skin Deep* and to further raise public awareness, environmental health activist and cofounder of the Campaign for Safe Cosmetics Stacy Malkan published a powerful exposé of the cosmetics industry in 2007. Her book, *Not Just a Pretty Face*,[58] reveals the truth about the cosmetics industry. It attracted national media coverage, won several awards and recommendations, and formed the basis for a documentary called *The Story of Cosmetics*,[59] codirected by Annie Leonard.

The Campaign for Safe Cosmetics complements its hardball tactics with a warmer, softer approach. By working directly with major manufacturers, it has persuaded some to reduce or eliminate toxics from their products. For instance, Revlon and L'Oreal have reformulated their products to meet European Union safety standards, which ban many of the most toxic chemicals from personal care products, and in 2012 Johnson & Johnson announced that it would remove carcinogens and other toxics from its baby and adult products. To add to this, from 2004 to 2011, the Campaign coordinated the Compact for Safe Cosmetics, a voluntary industry commitment to safety and transparency, with more than fifteen hundred signatory companies. After five years, the Campaign released a report highlighting more than three hundred "champion" cosmetics companies that met the Compact's goals.[60]

One of the challenges now facing the U.S. environmental health movement is the explosive growth of so-called natural products, including personal care and other consumer products. Hyped by the advertising industry as organic and healthy—their favorite words—it's next to impossible for consumers to distinguish between genuinely safer products and pretenders. With little help from government agencies, how can they know what to believe? So it's up to the U.S. environmental health movement and consumer organizations to provide the public with information about the products they buy, and to work with companies who want to make and sell safer products. And this is happening. In late 2012, Walgreens launched its Ology line of personal care, baby, and household cleaning products. With advice from Healthy Child, Healthy World, an environmental health group dedicated to protecting children from harmful chemicals, the company is offering safer, affordable products.

POLLUTION IN PEOPLE

It isn't just that everyone is exposed to toxics in consumer products, they're also inside our bodies, and the U.S. environmental health movement is using this fact to its advantage. Often called "pollution in people," this issue is the most definitive way to personalize environmental issues. The first study of pollution in people done by a nongovernmental organization was Barry Commoner's Baby Tooth Survey, which looked at levels of strontium 90 (see chapter 3). Since then, environmental health groups have done studies on toxic chemicals in a variety of human tissues, including hair, mothers' milk, urine, blood, and umbilical cord blood.

In 2002, Joe Thornton and his colleagues at the Environmental Working Group showed that more than 150 chemicals were present in nine people who had not had any unusual exposures;[61] in 2005, the Washington Toxics Coalition examined levels of six types of toxic chemicals in ten state residents;[62] in 2007, the Oregon Environmental Council did a similar study on ten Oregonians[63] and the Body Burden Work Group and Commonweal published the results of a seven-state study of thirty-five people;[64] and in 2010, the Collaborative on Health and the Environment's Learning and Developmental Disabilities Initiative released a report on levels of chemicals that are known or suspected to be neurotoxicants or are endocrine disruptors in twelve people.[65] The results of these studies are consistent with studies conducted by the Centers for Disease Control and Prevention.

Research on pollution in people drives home the reality of toxic contamination. Just as people with environmentally related diseases put a human face on environmental problems, so studies on pollution in people make it clear that everyone is contaminated: Everyone—from the newborn to the elderly and from people living in the Arctic to those living in Australia. Some of the most frequently detected substances include DDT, PCBs, PBDEs, and phthalates. These studies demonstrate the ubiquitous nature of toxic pollution and highlight the fact that our bodies are part of the environment. Revealing that toxics are not just "out there" in the environment, but that they are "in here" in the cells and tissues of our own bodies, these studies are extremely powerful. The realization that our bodies are just as, or even more, contaminated than the animals, plants, air, water, and soil of the earth is deeply shocking.

The U.S. environmental health movement has been very skillful in the way it has presented this information to the American public. By using phrases like "toxic trespass" and "chemical trespass," it has linked the presence of toxics in human tissues with deeply held American beliefs in private property and the sanctity of human life. This strategy is brilliant because it highlights the fact that our bodies—the ultimate form of private property—have been violated by toxic chemicals without our knowledge or consent. Talk about the right to know and prior informed consent (see chapter 4)—we don't even have the right to know about the pollutants in our own bodies, and we certainly haven't given our consent! To make matters worse, there's very little that ordinary people can do, except demand an end to environmental pollution. Eating organic and avoiding products containing toxic chemicals can help, but even the most conscientious consumers are still exposed.

Some groups working on children's environmental health have taken the link between pollution in people and environmental quality even further by pointing out that a mother's womb is the very first human environment. Using scientific studies showing that many toxics, such as mercury, can cross the placenta and that prenatal exposure can affect fetal growth and development, they've made an excellent case for reducing maternal exposures. This strategy speaks to people and organizations concerned about the "unborn child," including religious groups. Some supported the environmental health movement's efforts to limit mercury emissions from coal and oil-fired power plants. For example, in discussing mercury emissions and their health effects, the Evangelical Environmental Network web site says: "Christians are called by our Savior and Lord, Jesus Christ, to love our neighbors and do unto others as we would have them do unto us. We are thereby called to protect our most vulnerable populations, including unborn children."[66] And a 2011 letter from the U.S. Conference of Catholic Bishops to the EPA supporting proposed regulations on emissions of mercury and air toxics stated: "The Conference supports a national standard to reduce such pollution. Such standards should protect the health and welfare of all people, especially the most vulnerable members of our society, including unborn and other young children."[67] This made for an unlikely, but fascinating alliance between religious groups and environmental health groups that resulted in the adoption of new tougher regulations.

This chapter has examined how the U.S. environmental health movement personalizes environmental issues. Whether it's getting support from people affected by environmentally related disease, their caregivers or the organizations that represent them, drawing attention to children's environmental health or highlighting the presence of toxic chemicals in food, consumer products and our own bodies, this strategy has made the issue of toxic chemicals very meaningful and immediate to the American public. By focusing on people's personal lives and how they are affected by contaminants, it has transformed environmentalism.

In adopting this strategy, the U.S. environmental health movement recognizes that it's not enough to present scientific facts and figures. The language of science is not well understood by nonscientists, and more importantly, it's rarely sufficient to motivate the American public to take collective action. To engage ordinary people, the movement knows it must also talk about real exposures and real health effects in real people because this reaches people at a deep level in a way that scientific information, on its own, cannot. This isn't to say that the U.S. environmental health movement is antiscience. As noted in chapter 1, it must use science because this is Western culture's principal way of knowing. And science, its limitations, and the environmental health movement's use of precaution are the subject of the next chapter.

NOTES

1. World Health Organization. 2006. *Preventing disease through healthy environments: Towards an estimate of the environmental burden of disease*. Available at: http://www.who.int/quantifying_ehimpacts/publications/preventingdisease. Accessed October 31, 2012.

2. Coleen A. Boyle, Sheree Boulet, Laura A. Schieve, Robin A. Cohen, Stephen J. Blumberg, Marshalyn Yeargin-Allsopp, Susanna Visser, and Michael D. Kogan. Trends in the preva-

lence of developmental disabilities in U.S. children, 1997–2008. *Pediatrics* 127: 1034–1042 (2011).

3. Learning and developmental disabilities initiative working group. Available at: http://www.healthandenvironment.org/working_groups/learning. Accessed October 31, 2012.

4. Steven Gilbert, Elise Miller, Joyce Martin, Laura Abulafia. Scientific and policy statements on environmental agents associated with neurodevelopmental disorders. *Journal of Intellectual and Developmental Disability* 35(2): 121–128 (2010).

5. National Cancer Institute. Breast cancer risk in American women. Factsheet reviewed 09/24/2012. Available at: http://www.cancer.gov/cancertopics/factsheet/detection/probability-breast-cancer. Accessed October 31, 2012.

6. European Commission. *State of the art assessment of endocrine disrupters*. Available at: http://ec.europa.eu/environment/endocrine/documents/studies_en.htm. Accessed October 31 2012.

7. Anna M. Soto and Carlos Sonnenschein. Environmental causes of cancer: endocrine disruptors as carcinogens. *National Review of Endocrinology* 6: 363–370 (2010).

8. James T. Brophy, Margaret M. Keith, Andrew Watterson, Robert Park, Michael Gilbertson, Eleanor Maticka-Tyndale, Matthias Beck, Hakam Abu-Zahra, Kenneth Schneider, Abraham Reinhartz, Robert DeMatteo, and Isaac Luginaah. Breast cancer risk in relation to occupations with exposure to carcinogens and endocrine disruptors: a Canadian case-control study *Environmental Health* 11:87 (2012).

9. Janet Gray. *State of the evidence: The connection between breast cancer and the environment, sixth edition.* (2010). Breast Cancer Fund. Available at: http://www.breastcancerfund.org/assets/pdfs/publications/state-of-the-evidence-2010.pdf. Accessed October 31, 2012.

10. Breast Cancer and the Environment: Prioritizing Prevention. Report of the Interagency Breast Cancer and the Environmental Research Coordinating Committee (2013). Available at : http://www.niehs.nih.gov/ibcercc. Accessed February 13, 2013.

11. American Lung Association. *Trends in asthma morbidity and mortality*. September 2012. Available at: http://www.lung.org/finding-cures/our-research/trend-reports/asthma-trend-report.pdf. Accessed March 1, 2012.

12. Samuel Epstein. *The politics of cancer*. San Francisco, CA: Sierra Club Books (1978).

13. Samuel Epstein. *The politics of cancer revisited*. East Ridge Press (1998).

14. Devra Davis. *The secret history of the war on cancer*. New York: Basic Books (2009).

15. Canadian Cancer Society. *Cancer and the environment*. Available at: http://www.cancer.ca/Canada-wide/Prevention/Harmful%20substances%20and%20environmental%20risks.aspx?sc_lang=en. Accessed October 31, 2012.

16. Florence Nightingale. *Notes on nursing: What it is and what it is not*. Originally published in 1860. Dover Publications (1969).

17. Hollie Shaner-McRae, Glenn McRae, and Victoria Jas. Environmentally safe health care agencies; Nursing's responsibility, Nightingale's legacy. *Online Journal of Issues in Nursing* 12(2) (2007). Available at: http://www.nursingworld.org/MainMenuCategories/ANAMarketplace/ANAPeriodicals/OJIN/TableofContents/Volume122007/No2May07/Environmentally-SafeHealthCareAgencies.html. Accessed October 31, 2012.

18. Andrew M. Pope, Meta A. Snyder, and Lillian H. Mood (eds). *Committee on enhancing environmental health content in nursing practice, Institute of Medicine. Nursing, Health and the Environment*. Washington, DC: National Academies Press (1995).

19. American Nurses Association House of Delegates. Resolution on the reduction of health care production of toxic pollution. (1997). Available at: http://nursingworld.org/MainMenuCategories/WorkplaceSafety/Environmental-Health/PolicyIssues/ANAResolution_1_2.html. Accessed October 31, 2012.

20. Alliance of Nurses for Healthy Environments. Available at: http://e-commons.org/anhe. Accessed October 31, 2012.

21. Notes from the field: Congress on environmental health. *American Journal of Public Health* 54(3): 548–549 (1964).

22. Andrew M. Pope and David P. Rall (eds). *Environmental medicine: Integrating a missing element into medical education*. Institute of Medicine Report. Washington, DC: National Academy Press; 1995.

23. American Medical Association. H-135.973 Stewardship of the Environment. CSA Rep. G, I-89; Amended: CLRPD Rep. D, I-92; Amended: CSA Rep. 8, A-03.

24. American Medical Association. Report 4 of the Council on Scientific Affairs, Educational and Informational Strategies for Reducing Pesticide Risks (resolutions 403 and 404). (1994).

25. Health & Environment: A National Environmental Education Foundation Program. *Position statement on health professionals and environmental health education.* (2004). Available at: http://www.neefusa.org/pdf/PositionStatement.pdf. Accessed October 31, 2012.

26. Joseph S. Zickafoose, Stuart Greenberg, and Dorr G. Dearborn. Teaching home environmental health to resident physicians. *Public Health Reports* 126(Suppl. 1): 7–13 (2011).

27. Ruth Etzel (ed). *Pediatric environmental health*. 3rd Edition. American Academy of Pediatrics (2011).

28. Ruth A. Etzel, Ellen F. Crain, Benjamin A. Gitterman, Charles Oberg, Peter Scheidt, and Philip J. Landrigan. Pediatric environmental health competencies for specialists. *Ambulatory Pediatrics* 3(1): 60–63 (2003).

29. American College of Preventive Medicine. *ACPM statement on reform of the federal Toxic Substances Control Act of 1976*, January 8, 2012. Available at: http://c.ymcdn.com/sites/www.acpm.org/resource/resmgr/policyissues-files/acpm_tsca_statement.pdf. Accessed October 31 2012.

30. Eric Chivian, Michael McCally, Howard Hu, and Andrew Haines (eds). *Critical condition: Human health and the environment.* Boston, MA: MIT Press (1993).

31. Ted Schettler, Jill Stein, Fay Reich, and Maria Valente. *In harm's way: Toxic threats to child development.* A report of the Greater Boston Physicians for Social Responsibility. (2000). Available at: http://www.psr.org/site/PageServer?pagename=boston_inharmsway. Accessed October 31, 2012.

32. Jill Stein, Ted Schettler, Ben Rohrer, and Maria Valenti. *Environmental threats to healthy aging.* Greater Boston Physicians for Social Responsibility and the Science and Environmental Health Network. (2008). Available at: http://www.agehealthy.org/. Accessed October 31, 2012.

33. National Research Council Committee on Pesticides in the Diets of Infants and Children. *Pesticides in the diets of infants and children.* Washington, DC: National Academies Press (1993).

34. Natural Resources Defense Council. *Intolerable risk: Pesticides in our children's food.* Washington, DC (1989).

35. Children's Environmental Health Network. *Preventing child exposure to environmental hazards: Research and policy issues.* Symposium proceedings. Available at: http://www.cehn.org/files/Preventing_Child_Exposures_to.pdf. Accessed October 31, 2012.

36. Carol Browner. *Report on environmental health risks to children national agenda to protect children from environmental threats*. (November 1996).

37. Executive Order 13045—Executive Order on the protection of children from environmental health risks and safety risks. Available at: http://www.epa.gov/fedreg/eo/eo13045.htm. Accessed October 31, 2012.

38. Jill Stein, Ted Schettler, Ben Rohrer, and Maria Valenti. *Environmental threats to healthy aging.* Greater Boston Physicians for Social Responsibility and the Science and Environmental Health Network. (2008). Available at: http://www.agehealthy.org/. Accessed March 1, 2012.

39. Frances Moore Lappé. *Diet for a small planet*. Ballantine Books (1971).

40. U.S. Department of Agriculture, National Agricultural Statistical Service. *2007 Census of agriculture—state data.* Available at: http://www.agcensus.usda.gov/Publications/2007/Full_Report/Volume_1,_Chapter_2_U.S._State_Level/st99_2_044_044.pdf. Accessed October 31, 2012.

41. Steven McFadden. *The history of community supported agriculture, Part I: Community farms in the 21st century, poised for another wave of growth?* The Rodale Institute. Available

at: http://newfarm.rodaleinstitute.org/features/0104/csa-history/part1.shtml. Accessed October 31, 2012.

42. USDA. Farmers' markets and local food marketing. Available at: http://www.ams.usda.gov/AMSv1.0/ams.fetchTemplateData.do?template=TemplateS&left-Nav=WholesaleandFarmersMarkets&page=WFMFarmersMarketGrowth&descrip-tion=Farmers%20Market%20Growth&acct=frmrdirmkt. Accessed October 31, 2012.

43. Organic Trade Association. *2011 Organic industry survey*. Greenfield, MA: OTA.

44. Michael Pollan. *The omnivore's dilemma: A natural history of four meals*. New York: Penguin (2006).

45. Michael Pollan. *In defense of food: An eater's manifesto*. London, UK: Penguin Books (2008).

46. Michael Pollan. *Food rules: An eater's manual*. London, UK: Penguin Books (2009).

47. Michael Pollan. *Cooked: A natural history of transformation*. London, UK: Penguin Books (2013).

48. Eric Schlosser. *Fast food nation, the dark side of the all-American meal*. New York, NY: Houghton Mifflin (2001).

49. Marion Nestle. *Food politics*. Berkeley: University of California Press (2002).

50. Raj Patel. *Starved and stuffed: The hidden battle for the world food system*. Great Britain: Portobello Books (2007).

51. Michelle Obama. *American grown: The story of the White House kitchen garden and gardens across America*. Crown (2012).

52. Jessica Schifano, Joel Tickner, and Yve Torrie. *State leadership in formulating and reforming chemicals policy*. Lowell Center for Sustainable Production, University of Massachusetts Lowell. 2009. Available at: http://www.chemicalspolicy.org/downloads/StateLeadership_000.pdf. Accessed October 31, 2012.

53. National Environmental Trust. *Cabinet confidential: Toxic products in the home*. (2004). Washington, DC. Available at: http://www.csu.edu/CERC/documents/CabinetConfidential.pdf. Accessed October 31, 2012.

54. LOHAS (Lifestyles of Health and Sustainability). *About: LOHAS Background*. Available at: http://www.lohas.com/about. Accessed October 31, 2012.

55. Perry Romanowski and Randy Schueller. Fundamentals of cosmetic product safety testing. *Cosmetics and Toiletries* 111(10): 79(7) Oct. 1996.

56. Campaign for Safe Cosmetics, Research. Available at: http://safecosmetics.org/section.php?id=29. Accessed October 31, 2012.

57. Gregory Morris. Personal care suppliers invest in small expansions. *Chemical Week* May 11, 2009.

58. Stacy Malkan. *Not just a pretty face: The ugly side of the beauty industry*. Gabriola Island, BC, Canada: New Society Press (2007).

59. The Story of Cosmetics. Video. Available at: http://www.storyofstuff.org/movies-all/story-of-cosmetics/. Accessed October 31, 2012.

60. Campaign for Safe Cosmetics. *Market shift: The story of the compact for safe cosmetics and the growth in demand for safe cosmetics*. (2011). Available at: http://www.safecosmetics.org/downloads/MarketShift_CSC_Dec2011.pdf. Accessed October 31, 2012.

61. Joseph Thornton, Michael McCally, and Jane Houlihan. Biomonitoring of industrial pollutants: Health and policy implications of the chemical body burden. *Public Health Reports* 117: 315–323 (2002).

62. Washington Toxics Coalition. *Pollution in people: A study of toxic chemicals in Washingtonians*. (2005). Available at: http://pollutioninpeople.org. Accessed October 31, 2012.

63. Oregon Environmental Council. *Pollution in people*. (2007). Available at: http://www.oeconline.org/our-work/kidshealth/pollutioninpeople. Accessed October 31, 2012.

64. Body Burden Work Group & Commonweal Biomonitoring Resource Center. *Is it in us: Chemical contamination in our bodies*. (2007). Available at: http://www.isitinus.org/documents/Is%20It%20In%20Us%20Report.pdf. Accessed October 31, 2012.

65. The Learning and Developmental Disabilities Initiative. *Mind, disrupted: How toxic chemicals may change how we think and who we are*. (2010). Available at: http://

www.minddisrupted.org/documents/Mind%20Disrupted%20report.pdf. Accessed October 31, 2012.

66. Mercury & the unborn. Evangelical Environmental Network. Available at: http://creationcare.org/mercury/. Accessed October 31, 2012.

67. Letter to Lisa Jackson, Administrator of the EPA from the Most Reverent Stephen E. Blaire, Chairman, Committee on Domestic Justice and Human Development, U.S. Conference of Catholic Bishops .(June 20, 2011). Available at: http://www.usccb.org/about/general-counsel/rulemaking/upload/comments-to-epa-on-mercury-2011-06.pdf. Accessed October 31, 2012.

Chapter Seven

Precaution and the Limitations of Science

This chapter examines the limitations of science and how the U.S. environmental health movement has coped with them by advocating precaution. Advocating for precaution is now one of the principal strategies used by the U.S. environmental health movement. Asserting that it's morally wrong to postpone policy decisions because of a lack of complete scientific data, a precautionary approach is based on prudent foresight. It acknowledges that science and risk assessment—the scientific technique used to quantify the risks of exposure to hazardous substances, contaminated sites, and proposed development activities—are important tools for decision making, but it also recognizes their shortcomings.

Because Western science has been so effective at improving the quality of life, our culture has a naive belief it can always solve our problems—no matter their size or complexity. But this belief no longer serves us well. The issues we face today are much larger and more complex than ever before. The reductionist thinking that lies at the heart of Western science (see chapter 1) can provide "technical fixes"[1] that deal with specific, narrowly defined problems, but it can't address their root causes. For example, pollution control technologies can reduce environmental emissions, but they don't deal with the underlying problem of why pollution is generated in the first place. Fuel efficiency standards for automobiles can be improved, but this doesn't deal with the underlying problem of our dependence on cars. Buying "green" products can make us think we are doing something for the environment, but it doesn't deal with the underlying problem of the consumer society.

Scientific and technological solutions offer technical fixes that deal with presenting symptoms, rather than root causes. Although they are necessary, they aren't sufficient. Resolving the ecological crisis and preventing environmentally related diseases will require more systemic and holistic approaches that identify and deal with the root causes of the problem, not just presenting symptoms. And this will require changing our culture's values and beliefs, including anthropocentrism.

Science cannot provide much assistance with this work because of its built-in limitations. As described in the following pages, these limitations include the virtual impossibility of proving environmental causation, the failure to consider ethics, the distortion and cover-up of scientific information, and the problems associated with risk assessment.

THE IMPOSSIBILITY OF PROVING ENVIRONMENTAL CAUSATION

One of science's principal limitations is that it is based on the idea of proving causation—that something causes something else to happen. This distinguishes between chance associations of exposures and diseases, and when an exposure actually causes a disease. But there are two problems with trying to prove causation. First, science can only ever disprove a hypothesis (see chapter 1) and second the scientific bar for "proving" causation is just too high. Based on the scientific method, the criteria for proving environmental causation have evolved over the centuries and include:

- **Strength of Association:** There must be a strong association between the exposure and the observed health effect. For example, there's a strong association between smoking and lung cancer because the rate of lung cancer in smokers is about ten times higher than the rate in nonsmokers. But even so, it took decades before the science was considered strong enough to enact measures to control smoking. Few environmental exposures have anything like this strength of association; most are much weaker and so don't provide strong evidence of causation.
- **Specificity:** There must be a clear association between the exposure and the health effect. Researchers must be able to show that exposure to a specific environmental factor, and no other, caused the specific illness. There are two issues with this criterion: First, we are all exposed to many hazards, so who's to say that a particular exposure caused the observed illness? Second, most environmentally related diseases can be caused by many different factors, so who's to say that an illness was caused by an observed exposure? Scientists deal with this problem by using unexposed control groups or doing multivariant regression analyses that attempt to rule out the effects of "confounding factors" such as race, ethnicity, and economic status.

- **Biological Gradient:**[2] There must be a dose-response relationship between the exposure and the effect. In other words, scientists must show that low exposures cause fewer or less serious effects than higher doses. For instance, the fact that the death rate from lung cancer increases in proportion with the number of cigarettes smoked makes a case for causation. But scientists rarely know people's precise exposures to particular hazards because it's very difficult to measure them accurately.
- **Consistency:** The same health effect must be reported in different studies, conducted in different places, by different researchers at different times. If it is only seen in one study, or in one place, or at one time, or is reported by only one researcher, then the evidence for causation is weak. But is it ethical to let people get sick and die while scientists and regulators wait for enough studies to be done? And how many studies are enough, anyway?

Other criteria for proving causation include whether the cause precedes the effect (temporal relationship), whether the hypothesized cause and effect relationship is biologically plausible, whether the evidence fits with what is known regarding the natural history and biology of the outcome (coherence), whether there is experimental evidence, and whether the relationship can be reasoned by analogy. Even though these criteria have been around for a long time, they weren't formally articulated until 1965 when Austin Bradford Hill wrote his now classic essay called *The Environment and Disease: Association or Causation.*[3] Today, Bradford Hill's criteria remain the gold standard for demonstrating that something in the environment "causes" disease.

This burden of proof is much higher than any other in U.S. society. For instance, the legal system only requires a standard of "beyond a reasonable doubt." If toxic chemicals were "convicted" on the basis of the reasonable doubt standard, many more would have been banned or controlled. So why the double standard? Why is U.S. society willing to send people to jail or even administer the death penalty using the "beyond reasonable doubt" standard of evidence, but it's not willing to apply this standard to toxic chemicals that damage our health? It seems illogical and inconsistent.

Two main sciences have been used to try to satisfy the criteria for environmental causation: epidemiology and toxicology. Both rely on indirect methods because it is generally regarded as unethical to deliberately expose human beings to hazardous substances without their consent. These indirect methods include studying inadvertent, accidental exposures and occupational exposures in human beings (epidemiology) and studying the effects of hazardous substances on laboratory animals (toxicology), but both are fraught with difficulties.

One of the major challenges with environmental epidemiological studies is that determining peoples' exposures is very difficult, as mentioned above.

Exposures change over time and take place in many different settings: at home, work, school, in public places, and outside. No one knows precisely what they were exposed to, or how much, and it is difficult for researchers to keep track. Then there's the problem of specificity. It's often unclear that a disease was caused by a particular environmental agent. For instance, cancer can be caused by diet, lifestyle, or genetic factors, as well as toxic chemicals. So how can epidemiologists be sure precisely what's to blame?

A third major problem with many environmental epidemiological studies is that the strength of association between the exposure and the health effect is often weak. The lack of good exposure measures and the size of the excess risks associated with environmental exposures make it difficult to demonstrate strong associations. Related to this, many diseases, including some forms of cancer, do not develop until many years after the initial exposure. They have long latency periods, making it difficult to know exactly what caused them. For instance, the rare lung disease mesothelioma can take decades to become apparent after the onset of exposure to asbestos. In these circumstances, the strength of association is weak.

Because of these and other problems, most research on the health effects of toxic chemicals uses toxicological studies on laboratory animals or their tissues. Despite serious questions about the ethics of animal experimentation, toxicological studies are an important source of information on environmental health. Indeed, most government standards are based on toxicological studies.

But these studies come with their own challenges. Key among them is that translating health effects in laboratory animals to effects in human populations can never be completely accurate. There are just too many differences between animals and humans, differences in anatomy, physiology, metabolism, and exposures, to name a few. Moreover, in order to unequivocally identify health effects, most animal experiments use much higher doses than the public are normally exposed to. Therefore, to understand the likely effects in human populations, the results of animal studies must be extrapolated down to much lower exposure levels. But this extrapolation may not be accurate. The federal government has acknowledged many of these problems. In 2007, the National Research Council proposed a new approach to toxicity testing that would be more directly relevant to human exposures, rely less on animal experimentation, and be cheaper and faster,[4] and in 2009 the EPA developed a new strategic plan for evaluating chemical toxicity.[5] Meanwhile, in 2008 the federal government pooled its resources and created Tox21, a new screening method to predict chemical risks. Relying on mechanistic data and robotic technology, Tox21 is currently screening ten thousand substances of concern.

The problems with epidemiological and toxicological studies are widely recognized and make it virtually impossible to completely prove that something in the environment caused specific diseases or disabilities so a few epidemiologists are now talking about "the web of causation"[6] or recommending the use of systems diagrams to explain causation.[7, 8] But these considerations don't get sufficient attention from environmental health scientists or government regulators, and as a result, many environmental hazards go unregulated and environmental health problems continue to get worse. As long as government agencies require proof of causation before they act, very little will be done.

It's essential for the U.S. environmental health movement to deal with this problem; otherwise, it will never achieve its goal. But encouraged by corporate interests, scientists and legislators continue to insist on the impossibly high bar of causation. By failing to make allowances for the inherent limitations of science, they perpetuate a permissive approach to pollution.

THE FAILURE TO CONSIDER ETHICS

A second limitation is that science fails to consider ethics. It totally avoids questions about what **should** be done. Science doesn't tell us anything about what's right, fair, or just in particular situations. It can tell us what we are technologically capable of doing, but it's completely silent on the subject of what we **should** do. By simplifying environmental health issues into problems of cause and effect, science completely ignores ethical considerations. Indeed, the scientific method intentionally excludes them, proclaiming its objectivity. Because it aspires to be neutral and free of all values, science deliberately omits these critically important considerations.

But like it or not, values and ethics are always part of environmental health policy decisions. Should we build schools on top of toxic waste sites or not? Should we ban toxic chemicals that affect children's health or allow their continued use? Should we allow an industry to emit air pollutants into a minority community or not? These are ethical questions that science provides no help in answering. And, because legislators claim to rely on science, the values inherent in all environmental health decisions are rarely made explicit. Claiming to base their decisions on objective information, they often gloss over the ethical implications of their choices.

In contrast, the ancient Greeks understood the critical role that ethics play in decision making. Rather than trying to avoid the subject, they emphasized it. Aristotle called it "phronesis," which can be translated as practical wisdom. Requiring more than just technical knowledge or abstract ideas about good and bad, phronesis means acting in ways that encourage the full development of our humanity. But today, this way of thinking has been largely forgotten.

By downplaying the ethical implications of their environmental health decisions, legislators fail to consider critically important questions like:

- How can we do the most good and do the least harm in this situation?
- How can we respect the rights of everyone affected by this decision?
- How can we ensure fair and equal treatment for everyone involved?
- How can we best serve the entire community, not just a few individuals?
- Is this decision consistent with the sort of society we want?

These questions are vital for environmental health policy, but decision makers rarely address them head-on. And as a result environmental health policy often favors the power-holders and the privileged, while ignoring the consequences for poor and minority communities and other marginalized populations. The failure to consider ethics is an important limitation of science.

THE DISTORTION AND COVER-UP OF SCIENTIFIC INFORMATION

A third problem is that science is often distorted or covered up. Because environmental health policy places so much emphasis on scientific information, the methods used to collect, interpret, and communicate data must always be above suspicion. But over the years, there have been numerous instances where scientists, politicians, and corporate interests have deliberately distorted and hidden the scientific truth.

One of the most notorious examples was the Industrial Bio-Test (IBT) scandal. In the 1970s, IBT was the largest chemical testing laboratory in the United States, conducting about one-third of all toxicity testing in the country. But in 1983 the EPA announced that a 1976 audit had discovered "serious deficiencies and improprieties" in its toxicology studies, including "countless deaths of rats and mice that were not reported," "fabricated data tables," and "routine falsification of data."[9] The president of IBT, Dr. Joseph Calandra, and three of his senior associates were indicted for failing to follow standard scientific procedures, lying about the results, and then attempting to cover up what they had done. During the investigation, federal auditors found that IBT had falsified many of the test results used to register pesticides in the United States, Canada, Sweden, and other countries. Well over three hundred substances, including the fungicide captan, the herbicide silvex, and the insecticide toxaphene, were involved.[10, 11] The IBT executives were eventually found guilty and convicted, but the full environmental health effects of IBT's deception will never be known.

Less than a decade later, scientists were caught lying once again. In 1991, the EPA alleged that Craven Laboratories, a company that performed contract studies for pesticide companies including Monsanto, had falsified test results. Craven's deceptive practices included "falsifying laboratory notebook entries" and "manually manipulating scientific equipment to produce false reports." The following year, the owner and president and three employees were indicted on twenty felony counts. Several other employees agreed to plead guilty on a number of related charges. The owner was sentenced to five years in prison and fined $50,000, and Craven Labs was fined $15.5 million and ordered to pay $3.7 million in restitution.[12, 13]

In the 2000s, the distortion of scientific information was once again headline news, but in a very different way. This time, environmental scientists working for the federal government reported political censorship and interference by President George W. Bush's administration. Take the case of Dr. James Hansen, head of NASA's Goddard Institute for Space Studies and one of the world's leading experts on climate disruption. In a series of TV and newspaper interviews in 2005 and 2006, Hansen claimed that NASA administrators tried to edit his public statements about the causes of global warming after he criticized the federal government for failing to take action on the problem. In a story aired on CBS's *60 Minutes* TV show in 2006, Hansen stated that White House staff routinely rewrote press releases to make the science seem less threatening. He went on to say: "In my more than three decades in the government I've never witnessed such restrictions on the ability of scientists to communicate with the public."[14]

The censorship experienced by James Hansen is only one example of the political censorship and interference that plagued federal environmental science during the Bush administration. In 2008, the Union of Concerned Scientists (UCS) released the results of a national survey of almost fifty-five hundred EPA scientists. According to the results, "large numbers of EPA scientists reported widespread and inappropriate interference by EPA political appointees, the White House, and other federal agencies in their scientific work."[15] In fact, 60 percent of the survey respondents said they had experienced at least one incident of political interference between 2002 and 2007. In most cases, these incidents took the form of pressure to change scientific methods and results, editing of scientific documents by nonscientists, and delaying the release of scientific studies.

But it isn't just only fraudulent toxicity testing and political interference that threaten the integrity of environmental health science. It's also the way that scientific information is blatantly disregarded or manipulated by corporate interests. In fact, corporate interests have a habit, if not an addiction, to hiding and covering up science for the sole purpose of avoiding government regulation. One of the first books to expose these tactics was *Deceit and Denial*,[16] by Gerald Markowitz and David Rosner. Published in 2002, it provides a historical account of how the lead, plastics, and petroleum industries misled and lied to workers, consumers, and the general public about the health effects of their products.

One of the most egregious examples is the chemical industry's conspiracy to keep silent about vinyl chloride monomer—the chemical building block for PVC—and its potentially deadly health effects. By the early 1960s, industry studies showed that vinyl chloride monomer caused liver damage in laboratory animals, and by the late 1960s, liver cancers, a loss of bone in the fingers, and other health problems were showing up in workers at PVC plants. But despite plenty of evidence, the chemical industry continued to expose workers and said nothing. Instead, it deliberately covered up its findings and tried to keep them secret. Over a fifteen-year period, scientists were pressured to rewrite publications, manufacturers using PVC were kept in the dark, information was withheld from government health officials, studies were terminated to avoid producing damaging evidence, and workers were misled about the purpose of their health exams. A confidential 1964 B. F. Goodrich memo stated: "We would like to determine as quietly as possible whether similar disabilities might exist . . . and for this reason I would like to have you casually examine our employees' hands as part of any other medical service you provide to them." Continuing on, the memo states "I would appreciate your proceeding with this problem as rapidly as possible, but doing it incidentally to other examinations of our personnel. We do not wish to have this discussed at all, and I request that you maintain this information in confidence" (underlining in original).[17] Even after there was ample scientific evidence of vinyl chloride monomer's carcinogenicity, the chemical industry opposed the tightening of federal regulations. An internal 1974 Union Carbide memo stated the proposed Occupational Safety and Health Administration (OSHA) standard of no detectable level was "not necessary."[18]

More recently, the chemical companies that manufacture flame retardants used in furniture and other common household items were caught distorting scientific studies by the Chicago Tribune newspaper.[19] In 2012, it ran a series of articles revealing that industry claims about the effectiveness of its products were completely false. Not only has the industry downplayed scientific studies showing that some flame retardants can cause developmental effects in children and disrupt thyroid function, it has also attempted to avoid regulation by alleging that flame retardants give people a fifteenfold increase in time to escape fires. But they actually confer little or no protection. Even the researcher who conducted the original study said, "Industry has used this study in ways that are improper and untruthful."[20] In fact, scientists at the Consumer Product Safety Commission determined that the flame retardants in household furniture aren't effective. But the story doesn't end there. A noted burn surgeon, who testified before California lawmakers about the death of a seven-week-old baby who was allegedly burned in a fire while she lay on a pillow that lacked flame retardant chemicals, was lying. It turned out that there was no such baby and that his testimony was bought and paid for by an industry front group. This example reveals the tactics used by some corporate interests to avoid the regulation of their products.

The deliberate distortion and cover-up of scientific information and consequent lack of regulations have many consequences—all of them negative. First and foremost, they endanger the health and safety of countless numbers of people. If corporations evade regulation or if regulations are based on fraudulent or manipulated science, millions could suffer unnecessary environmentally related diseases and disabilities. Not only does this jeopardize the health of everyone alive today, it also threatens the health of future generations.

Second, distorting and covering up scientific information is fundamentally unethical and damages our social institutions. If people can't trust corporations to tell the truth and disclose what they know, they are less likely to buy their products or invest in them. This can be extremely damaging and harm corporate interests over the long term. But more importantly, when politicians distort science, they undermine the very basis of the social contract between a government and its citizens. If we can't rely on the government to tell the truth and protect our health, who can we count on? By weakening confidence in the structures of governance, political interference erodes the credibility of our administrative agencies, demoralizes the public, and threatens the democratic system of government.

PROBLEMS WITH RISK ASSESSMENT

Of all the limitations of science, the one that has preoccupied the U.S. environmental health movement is risk assessment. Risk assessment is a scientific tool that attempts to quantify the health risks of proposed developments, exposure to hazardous substances, and contaminated sites. Using extremely complex mathematical models, risk assessments are usually incomprehensible to nonscientists. Because of this, they make it difficult for ordinary people to understand the magnitude of environmental health problems. They also make it difficult for the public to participate fully in policy decisions, thereby limiting their democratic rights.

The method of risk assessment was originally developed by the Food and Drug Administration in the 1930s to set allowable levels for food additives, and it now provides the scientific rationale for most environmental health policies and regulations. Indeed, it has become an indispensable part of environmental health decision making. Today, government risk assessments comprise the following steps:[21]

- Hazard identification;

- Dose-response assessment;
- Risk characterization; and
- Exposure assessment.

The EPA conducted its first risk assessment in 1975.[22] Less than six months later, the Agency signaled its intent to adopt it as a policy-making tool by issuing *Interim Procedures and Guidelines for Health Risk and Economic Impact Assessments of Suspected Carcinogens*, which stated that "rigorous assessments of health risk and economic impact will be undertaken as part of the regulatory process."[23] But it wasn't until 1983 that the National Research Council published its *Red Book*, which contained a framework for the process.[24] In 1984, the EPA adopted this framework as the basis for a more detailed process.[25] Later, the Agency published a series of guidelines for conducting risk assessments for carcinogenic chemicals,[26] mutagens,[27] and chemical mixtures.[28] Many other federal agencies followed the EPA's lead and adopted risk assessment, including the Consumer Product Safety Commission, the Nuclear Regulatory Commission, the Occupational Safety and Health Administration, and the U.S. Departments of Agriculture, Defense, and Energy.

In the 1970s and 1980s, most environmentalists welcomed the introduction of risk assessment. A few questioned the idea of setting "acceptable risk" levels, but many saw the process as a way to speed up the regulatory process and make it more transparent. Most environmental groups believed that by using a standardized approach and a common measurement system regulations would be enacted more quickly and would be less contentious. But exactly the opposite has happened. As the process has become more complicated, it has become increasingly controversial and time-consuming. Instead of expediting regulations to protect environmental health, it has become a massive delaying tactic.

Take dioxins, for example. The EPA released its first risk assessment of this group of toxic substances in 1985, calling them probable human carcinogens.[29] But in 1991, EPA administrator William Reilly bowed to industry criticism and announced that a reassessment would be conducted. In 1994, the Agency released the draft reassessment, which reaffirmed the results of the original report and provided new evidence of harm.[30] In 1995, a decade after publication of the original document, the EPA's Science Advisory Board (SAB) reviewed the draft reassessment and requested major revisions.[31] After considerable additional work, the EPA issued a final draft report in 2000. The EPA SAB reviewed this document and issued its recommendations in 2001.[32] Then in 2003, the agency announced it would send the final draft to the National Academy of Sciences (NAS) for review. Some three years later, the National Academy of Sciences, which had been renamed the National Academies, released its comments.[33] Then in 2009, the

EPA held a public scientific workshop to help it respond to the National Academies report. Over a year later, in May 2010, the EPA released a draft report called *Reanalysis of Key Issues Related to Dioxin Toxicity and Response to NAS Comments*.[34] But it took until February 2012, more than twenty-five years after the initial risk assessment, for the EPA to release its final report.[35] And even then, it only released the first volume. It's unclear when the second volume will be published.

Although not all risk assessments take as long as this, many take years. This means it would take several millennia to assess the environmental health risks of each one of the approximately eighty thousand chemicals in use in the United States today, let alone the risks of the one thousand or so new substances entering the marketplace every year. The glacial speed of risk assessment has slowed the control of toxic chemicals and sabotaged environmentalists' well-meaning intentions to expedite regulation. And although the new Tox21 screening approach is being used to identify priority substances, the process still crawls along at a snail's pace.

Today, complete risk assessments have been done for only a handful of chemicals. Little or nothing is known about the health effects of most substances. In 1997, Environmental Defense found that even basic toxicity information was not available for nearly 75 percent of the top volume chemicals in commercial use in the United States.[36] And the situation hasn't improved significantly since then. Commenting on the implications of this lack of knowledge, a 2010 report prepared by the President's Cancer Panel said, "With nearly 80,000 chemicals on the market in the United States, many of which are used by millions of Americans in their daily lives and are un- or understudied and largely unregulated, exposure to potential environmental carcinogens is widespread."[37]

Unlike the environmental movement, many in the U.S. environmental health movement have criticized risk assessment. As well as arguing that it delays regulation, environmental health groups have identified many other problems, including:[38]

- **Inability to consider cumulative impacts:** With its origins in scientific reductionism, risk assessment cannot take account of the cumulative impacts of multiple environmental hazards. Even though we are exposed to many environmental hazards, risk assessment can only deal with one thing at a time. Hence, it provides little or no information about cumulative impacts. From the scant information that is available, we know that the effects of different environmental hazards can be synergistic; in other words, the combined effects can be greater than the sum of the individual effects. For instance, the combined health effects—in this case lung cancer—of exposure to radon gas and cigarette smoking are much greater than would be expected by simply adding together their separate effects.

This concern has been highlighted by the environmental justice movement (see chapter 8).

- **The illusion of certainty:** Risk assessment gives an illusion of certainty. Its use of numbers conveys confidence and sureness about the results. But in reality, risk assessment can never be completely precise or accurate. There are just too many unknowns, such as socioeconomic factors or gene-environment interactions. Uncertainty is an inherent part of risk assessment. Government agencies attempt to deal with this by including so-called "safety" or "uncertainty" factors in environmental health regulations, but although they are better than nothing, these factors can perpetuate a false sense of security.
- **Innocent until proven guilty:** Risk assessment assumes that substances and activities are innocent until proven guilty; that they are harmless until there is scientific evidence to the contrary. Because it is based on identifying hazards and then subjecting them to further study, risk assessment makes an *a priori* assumption of safety. This assumption has repeatedly been proven wrong. For instance, almost everyone believed that DDT and other organochlorine pesticides were safe when they were first introduced, but we now know this isn't true. To prevent environmentally related disease and disability, regulators should assume that substances and development activities could cause harm, unless they are proven safe.
- **Failure to protect the health of vulnerable individuals:** Risk assessment assumes that everyone responds to environmental exposures in the same way. But in reality, there is a huge variability in peoples' sensitivity. People with multiple chemical sensitivity (MCS) and children can experience health problems at very low exposure levels, while others may tolerate higher exposures. Hence, standards that are intended to protect "the average person" will not protect the health of vulnerable individuals. One size does not fit all.
- **Risk acceptability:** Risk assessment requires legislators to make decisions about acceptable levels of risk. This is based on the assumption that it's OK to harm people's health up to a certain "acceptable" level. This approach ignores the ethics of exposure and risk. Ethical environmental health policies are based on reducing or preventing harm, rather than risk acceptability. Moreover, decisions about risk acceptability are usually made by government agencies behind closed doors and with little or no input from the people who will be affected.
- **Failure to consider environmental justice:** Risk assessment doesn't consider environmental justice. It ignores questions about who will benefit from the substance or development activity and who must live with the risks. In our profit-driven society, it's often corporations who benefit and the public who must live with the risks. Even worse, it's often low-income

communities, people of color, and children who face the greatest health risks from environmental hazards (see chapter 8).

- **Reduces everything to numbers:** Risk assessment assumes that all exposures and effects can be reduced to numbers. In other words, it assumes that everything can be measured and quantified. But, as discussed in the Introduction, how can you measure the pain and suffering caused by environmentally related diseases and disabilities? How can you measure the emotional and psychological distress experienced by the victims, let alone the consequences for their families, friends, and communities? No one can put a number on these things. With its reliance on mathematical models and statistics, risk assessment just ignores the human costs of environmental pollution.

The National Research Council has acknowledged some of these problems, and in 2008 and 2009, it published two reports recommending modernizing the U.S. risk assessment system.[39, 40] In a moment of unusual candor, the 2009 report—known as the Silver Book—declared "the Red Book framework was not oriented to identifying the optimal process for complex decision-making." This is tantamount to saying that the 1983 framework has failed as a tool for making policy. This is a remarkable admission that would not have been made without the U.S. environmental health movement's critique of risk assessment. However, the environmental health movement hasn't just opposed risk assessment; it has suggested ways of strengthening it[41] and proposed a positive solution—precaution.

OVERVIEW OF PRECAUTION

According to the online Merriam Webster dictionary, precaution is "care taken in advance" and "a measure taken beforehand to prevent harm or secure good."[42] Its essence is captured in many age-old aphorisms, such as "an ounce of prevention is worth a pound of cure," "better safe than sorry," and "look before you leap." Precaution is a commonsense response to situations that could be hazardous, **before** they actually cause significant harm. It weighs the probabilities and comes down on the side of caution. Think about it this way: Would you get into an airplane if there was a 10 percent chance that it would crash and you'd be killed? Of course not. No one in their right mind would take these odds. Then why, as a society, do we gamble on toxics when the risks are much greater? Why do we take the chance? It just doesn't make sense. Unlike risk assessment, which is a scientific tool for environmental health decision making, precaution is a principle and an approach that can be broadly applied.

The best-known definition of the precautionary principle was developed in 1998 at a national meeting of environmental health leaders. Named after the location of the meeting—the Wingspread Conference Center in Racine, Wisconsin—the Wingspread Statement reads: "When an activity raises threats of harm to human health or the environment, precautionary measures should be taken even if some cause and effect relationships are not fully established scientifically. In this context the proponent of an activity, rather than the public, should bear the burden of proof. The process of applying the precautionary principle must be open, informed and democratic and must include potentially affected parties. It must also involve an examination of the full range of alternatives, including no action."[43] This definition has guided the environmental health movement ever since.

Precautionary approaches have been applied to environmental health problems for centuries. Perhaps the best-known historical example of its use is the removal of the handle of the Broad Street water pump to prevent the spread of cholera in London in 1854 (see chapter 1). But despite this history, the concept of precaution was not articulated in policy until the early 1970s. Some scholars say it originated in Sweden and others in Germany. In Germany, the precautionary principle ("Vorsorgeprinzip") can be traced back to the first draft of a bill (1970) aimed at securing clean air. The most unambiguous elaboration of the idea is from slightly later and reads: "Responsibility towards future generations commands that the natural foundations of life are preserved and that irreversible types of damage, such as the decline of forests, must be avoided."[44]

Since then, the precautionary principle has been included in many international agreements and declarations, including the Declaration of the International Conference on the Protection of the North Sea (1984),[45] the Montreal Protocol on Substances that Deplete the Ozone Layer (1987),[46] the Bergen Declaration on Sustainable Development (1990),[47] the Maastricht Treaty on European Union (1992),[48] the UN Framework Convention on Climate Change Convention (1992),[49] the Biosafety Protocol of the Convention on Biological Diversity (2000),[50] and the Stockholm Convention on Persistent Organic Pollutants (2001).[51] Most notably, the Rio Declaration signed at the conclusion of the 1992 United Nations Conference on Environment and Development states: "In order to protect the environment, the precautionary approach shall be widely applied by States according to their capabilities. Where there are threats of serious or irreversible damage, lack of full scientific certainty shall not be used as a reason for postponing cost-effective measures to prevent environmental degradation."[52] The precautionary principle is also the basis of REACH, the new approach to chemicals management introduced in Europe in 2007 (see chapter 5), and has been included in some national legislation, such as the *Canadian Environmental Protection Act* (1999).[53]

In the United States, however, the need for precaution has not been mentioned in any federal legislation. By 1998, this spurred environmental health activists into action. In that year, the Science and Environmental Health Network (SEHN) convened a national meeting of the movement's leaders to discuss how the precautionary principle could be advanced in the United States. Its participants called for government, corporations, communities, and scientists to implement precaution in environmental health policies, and developed the definition quoted on the previous page. In the years since the Wingspread meeting, SEHN has been a leading proponent of the precautionary principle in the United States.

Although SEHN and other environmental health groups have not succeeded in getting a commitment to precaution into federal legislation, they have raised public awareness. In December 2001, the New York Times described the precautionary principle as one of the most influential ideas of the year saying, "the precautionary principle poses a radical challenge to business as usual in a modern, capitalist, technological civilization. . . . If introduced into American law, the precautionary principle would fundamentally shift the burden of proof. . . . For the precautionary principle recognizes the limitations of science—and the fact that scientific uncertainty is an unavoidable breach into which ordinary citizens sometimes must step and act."[54] But what does precaution actually mean for environmental health and what are its ingredients?

THE INGREDIENTS OF PRECAUTION

Based on work conducted by supporters of the environmental health movement,[55, 56, 57, 58] the basic ingredients of precaution are:

- **Prevent Harm in the Face of Scientific Uncertainty:** Precaution is a special case of prevention—one of the core values of public health. Unlike prevention, which can be defined as taking action when there is unequivocal evidence of harm, precaution is about taking action when the scientific information is incomplete. For instance, requiring people to wear car seat belts was a preventive measure because there was irrefutable evidence that it reduced injuries and fatalities. In contrast, "virtual elimination" policies for toxics are precautionary because some of the cause and effect relationships between toxic exposures and health effects have not yet been fully established. But despite this difference, both are based on the idea that avoiding illness is better than treating it after the fact.

  Unfortunately, there are all too many examples of when the failure to take precautionary action has resulted in real health problems. In a 2001 study, called *Late Lessons from Early Warnings*, the European Environment Agency examined some of the worst examples, including radiation, benzene, asbestos, and antimicrobial agents.[59] In all of these cases, early warning signs of harm were ignored because there wasn't conclusive scientific proof, and legislators chose to assume they were safe. They were wrong, and their inaction led to serious health consequences.

- **Set Social Goals:** Precaution encourages planning based on social goals. It asks, "What type of world do we want and how can we get there?" and "Will this substance or activity help us to create the society we want?" In the United States, these questions are rarely asked. This results in a reactive approach to environmental health policy.

  By raising questions about our collective hopes and aspirations, precaution reframes the discussion from an analysis of risks to the creation of possibilities. Based on the belief that if we set social goals, we will find ways to achieve them, precaution encourages scientific and technological innovation.
- **Assess and Use Alternatives:** Precaution focuses on assessing and using alternatives that will prevent harm, rather than deciding how much risk is acceptable. By encouraging the identification and use of the safest alternatives, it also stimulates scientific and technological innovation.

  Alternatives assessments require standardized criteria to determine which options are best, and this is the subject of some debate in the U.S. environmental health movement. What should the criteria be? How should they be applied? The GreenScreen™ for Safer Chemicals attempts to respond to these questions. Providing a chemical hazard assessment system, it can be used to identify safer alternatives and chemicals of high concern.
- **Reverse the Burden of Proof:** Reversing the burden of proof means making corporations responsible for the safety of their substances and activities, instead of relying on the public to prove them harmful. It says, "If you want to manufacture or sell something, then it is your responsibility to test it and show it is safe." This is the opposite of what happens now, where substances and activities are considered safe unless there is evidence of harm. For too long, industry has received the benefit of the doubt.
- **Ensure Democratic Participation:** Democratic participation is essential for precaution. Preventing harm, setting social goals, and assessing alternatives all involve decisions that should not be left up to politicians, government agencies, or scientists. If the purpose of environmental health policy is to protect the public's health, then everyone has a right to participate in making policy decisions. Unless the public is part of the process, environmental health risks will be imposed on us. We, the public, should have a meaningful role in the process because we are the ones living with the risks. We are the ones exposed to environmental hazards and suffering the health consequences.

Moreover, democratic participation results in better policy decisions. Ordinary people usually think more broadly and systemically than scientists and politicians because they are not bound by disciplinary or political constraints. As well, they often have relevant and important information. As a result, the public often sees problems, issues, and solutions that the so-called experts miss.

Over the years, these ingredients have been used to develop numerous applications of the precautionary principle. Some are described in the next section.

PROGRESS ON PRECAUTION

Since the Wingspread meeting, the U.S. environmental health movement has lobbied government agencies to incorporate precaution into legislation and policies. And it has had some success. Between 1999 and 2010, at least nine states, cities, counties, and their agencies enacted precautionary policies, including:

- **The Los Angeles Unified School District's Policy on Integrated Pest Management:** The Los Angeles Unified School District, the second largest school district in the country, featured a commitment to precaution in its 1999 Policy on Integrated Pest Management (IPM).[60]
- **San Francisco Board of Supervisors Policies:** In 2003, the Board of Supervisors of the City and County of San Francisco became the first local government in the United States to make the precautionary principle the basis for all its environmental policy.[61] Subsequently, in 2005, the Board unanimously passed the Precautionary Purchasing Ordinance, which established goals and procedures for environmentally preferable purchasing by City departments.[62]
- **Hawaii Senate and House Resolutions:** In 2004, Hawaii's Senate and House adopted resolutions urging state agencies to implement the precautionary principle.[63] The precautionary principle has also been invoked by Hawaii's Supreme Court, which stated: "Where scientific evidence is preliminary and not yet conclusive regarding the management of fresh water resources which are part of the public trust, it is prudent to adopt 'precautionary principles' in protecting the resource."[64]
- **Toxics Reduction Strategy in Multnomah County:** In 2006, the Multnomah County Board of Commissioners in Oregon passed a resolution adopting a Toxics Reduction Strategy based on the precautionary principle.[65]
- **Comprehensive Plan in Seattle:** The City of Seattle included the precautionary principle in the Environment Element of its 2006 Comprehensive

Plan, saying, "The City will continue to engage the community about ways in which the City can give consideration to the precautionary principle."[66]

- **Precautionary Principle Ordinance in Lyndhurst:** In 2008, the town of Lyndhurst, New Jersey, adopted a precautionary principle ordinance (2674). Section 22–8.2 reads, "The Township of Lyndhurst will utilize the Precautionary Principle to develop laws for a healthier environment. By doing so, the Township will create and maintain a healthy, viable environment for current and future generations, and will become a model of sustainability."[67]

Environmental health groups and their supporters played a critical role in all of these examples, often forming *ad hoc* coalitions and partnerships. For instance, in Los Angeles, the California Safe Schools Coalition, comprising environmentalists, parents, doctors, principals, teachers, food services directors, school board members, and others, was instrumental in gaining support for the IPM policy. Similarly, in Seattle, the Precautionary Principle Working Group of the Collaborative on Health and Environment Washington had representatives from many different sectors and prepared extensive background materials for local councilors.

In the past few years, the U.S. environmental health movement has broadened its strategy to advance precaution. As well as advocating for government policies, it is also promoting corporate endorsements and applications. In fact, several large transnational companies are using precautionary approaches in their manufacturing processes, including the Body Shop, Bristol-Myers Squibb, Dell, Nike, and Samsung. For example, in its 2009 *Living Our Values* report, the Body Shop stated, "We adopt a precautionary approach to the use of chemicals,"[68] and in 2011 Nike stated: "In support of the principles of prevention and precaution . . . NIKE, Inc. supports the goal of systemic change to achieve zero discharge of hazardous chemicals associated with supply chains and the lifecycles of products within one generation or less."[69] Precautionary thinking is influencing many more companies, such as the normally conservative Ford Motor Company. The Company's 2010/11 Sustainability Report states: "We do not formally apply the precautionary principle to decision making across all of our activities. However, it has influenced our thinking. For example, in addressing climate disruption as a business issue, we have employed this principle."[70] Even the International Standards Organization's (ISO) 2600 voluntary international standard on social responsibility states that organizations should promote the precautionary approach.[71]

A 2009 study conducted for the Science and Environmental Health Network found that the concept of precaution has been adopted across a wide range of industries, disciplines, and movements.[72] Describing examples from agriculture, arts and culture, construction, food services, forestry, health care, and other sectors, the study concluded that the four most important factors in determining whether precautionary strategies are used are whether the concept has been institutionalized in decision-making processes, the capacity to envision what precaution looks like, the identification of measures to assess progress towards achieving precaution, and economic drivers for precaution.

The U.S. environmental health movement is unanimous that precaution is essential to prevent environmentally related diseases. Viewing it as an overarching principle and a practical approach that can incorporate risk assessment, the movement is fully committed to precaution. In contrast, the opponents of precaution, such as the American Chemistry Council, argue that precaution is fundamentally incompatible with risk assessment. Going further, they attack the idea, claiming it is antiscience and impedes scientific research and technological innovation. These and other criticisms have been convincingly refuted by environmental health activists who contend the opposite: that precaution deals with the limitations inherent in risk assessment and stimulates scientific research and technological innovation.[73, 74, 75]

Some of the strongest supporters of precaution are environmental justice activists. Alarmed that risk assessment has done nothing to alleviate the disproportionate burden of environmentally related disease experienced by people living in poverty, minority populations, and other marginalized communities, they argue that precaution is essential for environmental justice. Indeed, precaution is necessary to achieve both environmental health and environmental justice. Recognizing this connection between environmental health and environmental justice, the next chapter examines environmental justice and the right to a healthy environment.

NOTES

1. The phrase "technical fix" is taken from Ronald Heifetz's work on adaptive challenges and technical problems. According to Heifetz, technical problems are those that can be "fixed" by applying existing knowledge, values, and behavior. On the other hand, adaptive challenges require new knowledge, values, and behavior. It should be noted that technical problems are not necessarily technological (Ronald Heifetz. *Leadership without easy answers.* Cambridge, MA: Bellknap Press 1994).

2. This criterion has been challenged by research showing that the timing of exposure can be more important than the dose. For instance, studies on endocrine disrupting chemicals have shown that when an exposure takes place can be more important than its magnitude (see chapter 4).

3. Austin Bradford Hill. The environment and disease: Association or causation? *Proceedings of the Royal Society of Medicine* 58: 295–300 (1965).

4. National Research Council. *Toxicity testing in the 21st century: A vision and a strategy.* Washington, DC: The National Academies Press (2007).

5. EPA, Office of the Science Advisor. *The U.S. Environmental Protection Agency's strategic plan for evaluating the toxicity of chemicals.* EPA 100/K-09/001 (March 2009). Available at: http://www.epa.gov/spc/toxicitytesting/docs/toxtest_strategy_032309.pdf. Accessed October 31, 2012.

6. Nancy Krieger. Epidemiology and the web of causation: Has anyone seen the spider? *Social Science and Medicine* 39(7): 887–903 (1994).

7. Sander Greenland, Judea Pearl, and James Robins. Causal diagrams for epidemiologic research. *Epidemiology* 10(1): 37–48 (1999).

8. Michael Joffe, Manoj Gambhir, Marc Chadeau-Hyam, and Paolo Vineis. Causal diagrams in systems epidemiology. Emerging Themes in Epidemiology 9:1 (2012).

9. U.S. Congress. House of Representatives. Committee on Government Operations. 1984. *Problems plague the Environmental Protection Agency's pesticide registration activities*. 63rd report, House Report 98-1147. Washington, DC: U.S. Government Printing Office.

10. Keith Shneider. Faking it: The case against Industrial Bio-Test Laboratories. *Amicus Journal* Natural Resources Defense Council. (Spring 1983).

11. Carl Jensen. *20 years of censored news*. New York: Seven Stories Press (1995).

12. U.S. Environmental Protection Agency. Communications, Education, and Public Affairs. Press advisory. Craven Laboratories, owner, and 14 employees sentenced for falsifying pesticide tests. Washington, DC (March 4, 1994).

13. U.S. Department of Justice, United States Attorney, Western District of Texas. Texas laboratory, its president, 3 employees indicted on 20 felony counts in connection with pesticide testing. Austin, TX. (September 29, 1992).

14. *Rewriting the science*. CBS *60 Minutes*. March 19, 2006. Available at: http://www.cbsnews.com/2100-18560_162-1415985.html. Accessed October 31, 2012.

15. Union of Concerned Scientists. *Interference at the EPA: Science and politics at the U.S. Environmental Protection Agency*. Cambridge, MA. (April 2008).

16. Gerald Markowitz and David Rosner. *Deceit and denial: The deadly politics of industrial pollution*. Berkeley, CA: University of California Press (2002).

17. Correspondence form Ray Wilson to J. Neuman, B. F. Goodrich Company, dated November 12, 1964. Available at: http://www.chemicalindustryarchives.org/dirtysecrets/vinyl/pdfs/BFG50759-1.pdf#page=1. Accessed October 31, 2012.

18. Internal Union Carbide Correspondence, dated May 30, 1974. Available at: http://www.chemicalindustryarchives.org/dirtysecrets/vinyl/pdfs/UCC008994.pdf#page=2. Accessed October 31, 2012.

19. Chicago Tribune Watchdog. Playing with fire. Series of four articles published May 6, 8, 9, and 10, 2012. Available at: http://media.apps.chicagotribune.com/flames/index.html. Accessed October 31, 2012.

20. As quoted in: Sam Roe and Patricia Callahan. Distorting science. *Chicago Tribune*, May 9, 2012. Available at: http://www.chicagotribune.com/news/watchdog/flames/ct-met-flames-science-20120509,0,5238451,full.story. Accessed October 31, 2012.

21. National Research Council Committee on Improving Risk Analysis Approaches Used by the U.S. EPA. *Science and decisions: Advancing risk assessment.* Washington, DC: National Academies Press (2009). p. 11.

22. Arnold M. Kuzmack and Robert E. McGaughy. *Quantitative risk assessment for community exposure to vinyl chloride.* EPA Office of Planning and Management and Office of Health and Ecological Effects. (December 1975).

23. Russell Train. *Interim procedures and guidelines for health risk and economic impact assessments of suspected carcinogens*. EPA Office of the Administrator. (May 1976).

24. National Research Council. *Risk assessment in the federal government: Managing the process*. Washington, DC: National Academy Press (1983).

25. Environmental Protection Agency. *Risk assessment and management: Framework for decision making*. EPA 600/9-85-002. Washington, DC (1984).

26. *Guidelines for carcinogen risk assessment*. U.S. Environmental Protection Agency, Risk Assessment Forum, Washington, DC (1986).

27. *Guidelines for mutagenicity risk assessment.* U.S. Environmental Protection Agency, Risk Assessment Forum, Washington, DC (1986).

28. *Guidelines for the health risk assessment of chemical mixtures.* U.S. Environmental Protection Agency, Risk Assessment Forum, Washington, DC (1986).

29. U.S. Environmental Protection Agency. *Health effects assessment document for polychlorinated dibenzo-p-dioxins*. Prepared by the Office of Health and Environmental Assess-

ment, Environmental Criteria and Assessment Office, Cincinnati, OH, for the Office of Emergency and Remedial Response, Washington, DC. EPA/600/8-84/014F. (1985).

30. U.S. Environmental Protection Agency. *Health assessment document for 2,3,7,8-tetrachlorodibenzo-p -dioxin (TCDD) and related compounds, external review draft*. Prepared by the Office of Health and Environmental Assessment, Office of Research and Development, Washington, DC. EPA/600/BP-92/001a, b, c. (1994).

31. U.S. Environmental Protection Agency Science Advisory Board. *Re-evaluating dioxin: Science Advisory Board's review of EPA's reassessment of dioxin and dioxin-like compounds*. Washington, DC (1995).

32. U.S. Environmental Protection Agency. *Dioxin reassessment: an SAB review of the Office of Research and Development's reassessment of dioxin*. EPA-SAB-EC-01-006. (2001).

33. Committee on EPA's Exposure and Human Health Reassessment of TCDD and Related Compounds, National Research Council. *Health risks from dioxin and related compounds: Evaluation of the EPA reassessment*. Washington, DC: National Academies (2006).

34. U.S. Environmental Protection Agency. *Reanalysis of key issues related to dioxin toxicity and response to NAS comments, external review draft*. Washington, DC. (May 2010). Available at: http://cfpub.epa.gov/ncea/cfm/recordisplay.cfm?deid=222203. Accessed October 31, 2012.

35. EPA. *Reanalysis of key issues related to dioxin toxicity and response to NAS comments, Volume 1*. Washington, DC. (February 2012). Available at: http://www.epa.gov/iris/supdocs/dioxinv1sup.pdf. Accessed October 31, 2012.

36. Environmental Defense Fund. *Toxic ignorance: The continuing absence of basic health testing for top-selling chemicals in the United States*. Washington, DC (1997).

37. President's Cancer Panel. *Reducing environmental cancer risk: What we can do now 2008–2009 Annual Report*. Washington, DC: U.S. Department of Health and Human Services (2010). Available at: http://deainfo.nci.nih.gov/advisory/pcp/annualreports/pcp08-09rpt/PCP_Report_08-09_508.pdf. Accessed October 31, 2012.

38. See, for example, Mary O'Brien. *Making better environmental decisions, an alternative to risk assessment*. Cambridge, MA: MIT Press (2000); Peter Montague. Getting beyond risk assessment. *Rachel's Democracy & Health News* Issue 846, Mar. 16, 2006; and Peter Montague. A letter to my friend who is a risk assessor. *Rachel's Democracy & Health News* Issue 920 Thursday, August 16, 2007.

39. National Research Council of the National Academies, Committee on the Health Risks of Phthalates. *Phthalates and cumulative risk assessment: The tasks ahead*. National Academies Press: Washington DC, 2008. Available at:
http://www.nap.edu/catalog.php?record_id=12528. Accessed October 31, 2012.

40. National Research Council Committee on Improving Risk Analysis Approaches Used by the U.S. EPA. *Science and decisions: Advancing risk assessment*. Washington, DC: National Academies Press (2009).

41. Sarah Janssen, Jennifer Sass, Ted Schettler, and Gina Solomon. *Strengthening toxic chemical risk assessments to protect human health*. Issue Paper. Natural Resources Defense Council and the Science and Environmental Health Network (February 2012). Available at: http://www.nrdc.org/health/files/strengthening-toxic-chemical-risk-assessments-report.pdf. Accessed October 31, 2012.

42. Merriam Webster Dictionary (online). Available at: http://www.merriam-webster.com/dictionary/precaution. Accessed October 31, 2012.

43. *The Wingspread consensus statement on the precautionary principle*. Science and Environmental Health Network (1998). Available at: http://www.sehn.org/wing.html. Accessed October 31, 2012.

44. As quoted in: World Commission on the Ethics of Scientific Knowledge and Technology. *The precautionary principle*. Paris, France: United Nations Science, Educational and Cultural Organization (2005).

45. Declaration of the International Conference on the Protection of the North Sea. Bremen, Germany (1984). Available at: http://www.ospar.org/html_documents/ospar/html/1nsc-1984-bremen_declaration.pdf. Accessed October 31, 2012.

46. The Montreal Protocol on Substances that Deplete the Ozone Layer (1987).

47. Bergen Ministerial Declaration on Sustainable Development in the ECE Region (1990).
48. Maastricht Treaty Article III-233 (1992).
49. UN Framework Convention on Climate Change. Article 3.3 (1992).
50. Cartagena Protocol on Biosafety to the Convention on Biological Diversity. Preamble. (2000).
51. Stockholm Convention on Persistent Organic Pollutants. Preamble. (2001).
52. Principle 15. Rio Declaration on Environment and Development. Report of the UN Conference on Environment and Development, Rio de Janeiro. June 3–14, 1992.
53. *Canadian Environmental Protection Act*. Preamble. (1999).
54. Michael Pollan. The year in ideas: A to Z: Precautionary principle. *New York Times*. December 9, 2001. Available at: http://www.nytimes.com/2001/12/09/magazine/the-year-in-ideas-a-to-z-precautionary-principle.html?scp=1&sq=precautionary+principle&st=nyt. Accessed October 31, 2012.
55. Joel Tickner, Carolyn Raffensperger, and Nancy Myers. *The precautionary principle in action: A handbook.* 1st Edition. Written for the Science and Environmental Health Network. (undated).
56. David Kriebel and Joel Tickner. Reenergizing public health through precaution. *American Journal of Public Health* 91(9):1351–1361 (2001).
57. Joel Tickner (ed). *Precaution: Environmental science and preventive public policy*. Washington, DC: Island Press (2003).
58. Nancy Myers. *Precautionary procedures: Tools of analysis and intention*. In: Precautionary tools for reshaping environmental policy. Edited by Nancy Myers and Carolyn Raffensperger. Cambridge, MA: MIT Press (2006).
59. European Environment Agency. *Late lessons from early warnings: The precautionary principle 1896–2000*. Environmental Issues Report No. 22. Luxembourg (2001).
60. Los Angeles Unified School District, unpublished policy, cited in *Rachel's Environment & Health News* Issue 684 (January 26, 2000). Available at: http://www.rachel.org/files/rachel/Rachels_Environment_Health_News_1687.pdf. Accessed October 31, 2012.
61. The text of San Francisco's policy statement is published in *Rachel's Environment & Health News* Issue 765 (March 20, 2003). Available at: http://www.rachel.org/files/rachel/Rachels_Environment_Health_News_2338.pdf. Accessed October 31, 2012.
62. Legislative Digest. Available at: http://sfenvironment.org/sites/default/files/policy/sfe_th_precautionarypurchasing_ord.pdf. Accessed October 31, 2012.
63. Resolutions Adopted by the Hawaii State Legislature Regular Session of 2004. Available at: http://hawaii.gov/lrb/legis04/reso04.pdf. Accessed October 31, 2012.
64. As quoted by Carolyn Raffensperger in a presentation on The Precautionary Principle and the Public Trust Doctrine to the Los Angeles Air Quality Management District. (August 2003). Available at: http://www.aqmd.gov/ej/events/precautionary_principle/raffensperger.pdf. Accessed October 31, 2012.
65. Multnomah County, OR. *Toxics reduction strategy* (Res. 06-073). Available at: http://web.multco.us/sustainability/documents/toxics-reduction-strategy-resolution-may-2006. Accessed October 31, 2012.
66. *Seattle comprehensive plan: Towards a sustainable Seattle, environment element.* Available at: http://www.seattle.gov/DPD/static/environment_element_LatestReleased_DPDP_021135.pdf. Accessed October 31, 2012.
67. Peter Montague. (2008). Town of Lyndhurst, N.J. adopts the precautionary principle. *Rachel's Precaution Reporter* Issue 168. Available at: http://www.precaution.org/lib/08/prn_lyndhurst_passes_pp_law.081111.htm. Accessed October 31, 2012.
68. The Body Shop. *Living our values, the Body Shop International PLC Values Report 2009*. p. 40. Available at: http://www.thebodyshop-usa.com/pdfs/values-campaigns/Values_report_lowres_v2.pdf. Accessed October 31, 2012.
69. Nike Inc. *Commitment on zero discharge of hazardous chemicals.* Available at: http://nikeinc.com/news/nike-inc-commitment-on-zero-discharge-of-hazardous-chemicals. Accessed October 31, 2012.
70. Ford Motor Company. *Sustainability report 2010/11*. Available at: http://corporate.ford.com/microsites/sustainability-report-2010-11/environment. Accessed October 31, 2012.

71. *International Standards Organization 2600 Social responsibility, core subject: The environment.* (May 2010). Geneva, Switzerland: ISO.

72. Patrice Sutton. *Advancing the precautionary agenda.* Prepared for the Science and Environmental Health Network. (February 2009). Available at: http://www.sehn.org/pdf/Advancing%20the%20Precautionary%20Agenda.pdf. Accessed October 31, 2012.

73. Peter Montague. Answering the critics of precaution Part 1. *Rachel's Democracy and Health News* Issue 789. (April 15, 2004). Available at: http://www.rachel.org/?q=en/node/6464. Accessed October 31, 2012.

74. Peter Montague. Answering the critics of precaution Part 2. *Rachel's Democracy and Health News* Issue 790. (April 29, 2004). Available at: http://www.rachel.org/?q=es/node/6463. Accessed October 31, 2012.

75. Joel Tickner, David Kriebel, and Sara Wright. A compass for health: Rethinking precaution and its role in science and public health. *International Journal of Epidemiology* 32: 489–492 (2003).

Chapter Eight

Environmental Justice and the Right to a Healthy Environment

Extending the concept of social justice to include the environment, environmental justice asserts that no one should suffer health problems because of their zip code, their income, race, or ethnicity and that everyone should have a meaningful voice in policy decisions that affect them. But what exactly is environmental justice? The answer to this question is not as straightforward as it might appear. This chapter examines environmental justice and the right to a healthy environment. By exploring these topics in depth, it reveals the critical role that environmental justice plays in protecting environmental health. Achieving environmental health for all will be impossible without environmeental justice.

PERSPECTIVES ON ENVIRONMENTAL JUSTICE

Environmental justice is a contested concept. On one hand, it's a lens through which to analyze economic development and the pollution problems it causes. Raising questions about who benefits and who pays the price, this view of environmental justice highlights issues of power and privilege.[1] Seen this way, environmental justice is part of the academic discipline of environmental sociology and a framework for ordinary people to understand their shared experiences. Although there is something of a division between grassroots environmental justice activists and academics,[2] both use the concept as a social critique.

Not surprisingly, this understanding of environmental justice is held by the environmental justice movement. As discussed in chapter 4, this movement grew out of the civil rights movement, when African Americans became alarmed about toxic dump sites in their communities. Quickly making connections with the environmental discrimination experienced by other racial and ethnic minorities, such as Native Americans, they created a movement for all people of color. Later, as it became clearer that poor, white communities were also disproportionately affected by pollution, the environmental justice movement took on a broader socioeconomic analysis based on income, as well as race and ethnicity. More recently, the concept of environmental justice has been applied even more inclusively to anyone deprived of their environmental rights[3] including women, children, and other vulnerable or disadvantaged populations. As a result, the concept of environmental justice has become much more expansive and the movement has attracted a wider and more diverse range of supporters.

The view of environmental justice as social critique is evident in the seminal writings that have guided and shaped the environmental justice movement. For instance, the 1991 *Principles of Environmental Justice* open with the words:

> **WE, THE PEOPLE OF COLOR**, gathered together at this multinational People of Color Environmental Leadership Summit, to begin to build a national and international movement of all peoples of color to fight the destruction and taking of our lands and communities, do hereby re-establish our spiritual interdependence to the sacredness of our Mother Earth; to respect and celebrate each of our cultures, languages and beliefs about the natural world and our roles in healing ourselves; to ensure environmental justice; to promote economic alternatives which would contribute to the development of environmentally safe livelihoods; and, to secure our political, economic and cultural liberation that has been denied for over 500 years of colonization and oppression, resulting in the poisoning of our communities and land and the genocide of our peoples, do affirm and adopt these Principles of Environmental Justice.[4]

Similarly, in his environmental justice framework, Robert Bullard, the father of environmental justice, states: "The environmental justice framework attempts to uncover the underlying assumptions that may contribute to and produce unequal protection. This framework brings to the surface the ethical and political questions of 'who gets what, why, and how much.'"[5]

On the other hand, environmental justice can be seen as the government policy principle that everyone should be treated equally and have significant involvement in decision-making processes. This perspective is evident in the EPA's definition of environmental justice: ". . . the fair treatment and meaningful involvement of all people regardless of race, color, national origin, or income with respect to the development, implementation, and enforcement of environmental laws, regulations, and policies. . . . It will be achieved when everyone enjoys the same degree of protection from environmental and health hazards and equal access to the decision-making process to have a healthy environment in which to live, learn, and work."[6]

This government view is sometimes conflated with the earlier concept of environmental equity, a softer and less politically charged term still used by some agencies that focuses on the equal distribution of risks, rather than righting a wrong or correcting an unjustly imposed burden. Although the term environmental justice is now used more frequently, government agencies use it in a narrower and more limited way than environmental justice activists. Emphasizing procedures to increase public participation in environmental decision-making processes and studies on the environmental health risks experienced by disadvantaged communities and populations, it's largely an administrative and scientific understanding of the phrase.

These two perspectives—social criticism and government policy principle—are connected because they both seek to reduce the disproportionate risks that are the hallmark of environmental injustice. As stated by Agyeman and Evans in an article on "Just Sustainability," "environmental justice may be viewed as having two distinct but inter-related dimensions. It is, predominantly at the local and activist level, a vocabulary for *political opportunity, mobilization and action*. At the same time, at the government level, it is a *policy principle* that no public action will disproportionately disadvantage any particular social group" (emphasis in original).[7] Whether you think of environmental justice as social criticism or as a government policy principle, at its heart lies the belief that everyone has a right to a healthy environment.

CONSTITUTIONAL AND LEGAL RIGHTS TO A HEALTHY ENVIRONMENT

The belief that everyone is entitled to live in a healthy environment has transformed public debate on the environment. Just as the environmental health movement changed the debate by personalizing environmental issues, so the environmental justice movement changed it by talking about human rights. And in doing so, the environmental justice movement followed in the footsteps of many successful rights-based movements in the past. For instance, in the eighteenth century, the antislavery movement asserted that the basic human right to freedom applied to everyone and then framed slavery as a violation of this right. In the nineteenth century, the suffrage movement argued that women had the right to vote because they were equal to men and then pointed out that they did not have this right. Later in the 1960s and 1970s, the civil rights movement made the case that African Americans should have equal rights with whites, but did not. By using a similar approach as these movements, environmental justice is using a well-known and effective strategy for progressive social change.

Legal and constitutional rights to a healthy environment have a long and complex history. Although the environmental justice movement originated from the antislavery and civil rights movements, the right to a healthy environment goes much further back to when the Roman Emperor Justinian the Great first provided people with the right to use natural resources. As well as spreading Christianity throughout what remained of the Roman Empire, Justinian believed that free access to nature's bounty was essential for human health and well-being. So in the second book of his legal code—the *Institutiones* [8]—he codified the public's "rights of commons" by stating that the air, water, fish, sea, and shoreline were common property for use by everyone. By the Middle Ages, Justinian's rights had been narrowed and referred mostly to access to public land for grazing animals, however, "the commons" is still a widely understood concept today.

In fact, in recent years the U.S. environmental health movement has used this idea to argue for the right to a healthy environment. Carolyn Raffensperger, executive director of the Science and Environmental Health Network (SEHN), has said that "the mother of all commons is, of course, the Earth itself" and that the rights of commons are essential to protect and promote environmental health.[9] Seeing water, air, land, and biodiversity as part of humankind's common heritage, SEHN and other groups argue against the privatization of these resources and for their protection.

As well as drawing on Justinian's "rights of commons," U.S. environmental health and justice groups can refer to international, constitutional, and legal rights to a healthy environment. Forty years ago, the Declaration of the UN Conference on the Human Environment (1972) held in Stockholm stated: "Man has the fundamental right to freedom, equality and adequate conditions of life, in an environment of a quality that permits a life of dignity and well being, and he bears a solemn responsibility to protect and improve the environment for present and future generations,"[10] and in 1994, a UN report concluded that everyone has the right to a safe and healthy environment.[11] This report included a remarkably strong set of principles to guarantee this right. Although the UN failed to act on the report's recommendations, it recognized the human right to clean water and sanitation in 2010.[12]

Although the right to a healthy environment has not been recognized by the UN, it has been acknowledged by many national governments. In fact, the constitutions and laws of more than sixty countries provide their citizens with this right, at least in theory. For instance, South Africa's Constitution states that "Everyone has the right (a) to an environment that is not harmful to their health or wellbeing; and (b) to have the environment protected, for the benefit of present and future generations, through reasonable legislative and other measures that (i) prevent pollution and ecological degradation; (ii) promote conservation; and (iii) secure ecologically sustainable development and use of natural resources while promoting justifiable economic and social development."[13] Similarly, Hungary's Constitution "recognizes and enforces the right of everyone to a healthy environment."[14] Other countries with similar constitutional provisions include Ecuador, Peru, Portugal, the Philippines, and South Korea. Even the constitution of the obscure African country of Burkina Faso guarantees its citizens the right to a healthy environment.[15]

Even though it's usually difficult or impossible for citizens to exercise these rights, it's still important to have them. Providing written statements of peoples' environmental entitlements, constitutions and laws can be used in efforts to hold governments and corporations accountable for their actions. In countries that have enshrined the right to a healthy environment, environmental health and justice advocates can make a stronger case than in countries that have not recognized this right.

Citizens of the United States do not have the right to a healthy environment. Twice, Congress has considered serious proposals to amend the Constitution and twice it's failed to take action. In 1968, Senator Gaylord Nelson (D) proposed an amendment which read: "Every person has the inalienable right to a decent environment. The United States and every state shall guarantee this right."[16] And in 1970, Representative Richard Ottinger (D) offered an amendment which would guarantee Americans the "right . . . to clean air, pure water, freedom from excessive and unnecessary noise, and the natural, scenic, historic, and esthetic qualities of their environment shall not be abridged."[17] Undaunted by the failure of these proposals, environmental groups sued in federal court, asserting that the right to a healthy environment should be recognized under the Due Process Clause of the Fifth and Fourteenth Amendments or within the scope of the Ninth Amendment. Like the attempted constitutional amendments, these suits failed, although some courts acknowledged the environment's importance and the future possibility of constitutional environmental protection.[18] The closest that this country comes to the right to a healthy environment is the *National Environmental Policy Act* (1969) which states that "it is the continuing responsibility of the Federal Government to . . . assure for all Americans safe, healthful, productive, and aesthetically and culturally pleasing surroundings."[19]

In contrast, several state constitutions provide their citizens with the right to a healthy environment, including Hawaii, Illinois, Massachusetts, Montana, and Pennsylvania. Pennsylvania's Constitution declares: "The people have a right to clean air, pure water, and to the preservation of the natural, scenic, historic and esthetic values of the environment."[20] Hawaii's affirms that "Each person has the right to a clean and healthful environment, as defined by laws relating to environmental quality, including control of pollution and conservation, protection and enhancement of natural resources."[21] And Illinois' declares that "Every person has the right to a healthful environment."[22] Although these rights don't have many legal teeth, they can be used to educate the public and raise awareness about the need for environmental health and justice.

Advocating that everyone has the same right to a healthy environment makes a lot of sense in the United States because this country was founded on the ideal of equality. The 1776 Declaration of Independence contains the familiar words "We hold these truths to be self-evident, that all men are created equal, that they are endowed by their Creator with certain unalienable Rights, that among these are Life, Liberty and the pursuit of Happiness." Subsequently, the 1789 Bill of Rights gave "the people" additional, more specific rights, including freedom of speech and religion, the right to bear arms, and protection from unreasonable search and seizure. Even though these rights were only given to white men—women and slaves could not exercise them—the concept of human rights and equality is still part of this country's national identity.

But the right to a healthy environment isn't an easy sell in the United States. The belief in individual rights, and especially individual property rights, often makes it difficult to persuade people that there is a collective right to a healthy environment. This is because the concept of individualism is very deeply embedded in American culture. In contrast to many Indigenous cultures, which view the community or the group as the basic unit of society, American culture regards the individual as more important. The individual and her/his rights almost always come first. Just think of the right to bear arms, intellectual property rights, and the right to free speech. These and other similar rights reveal just how firmly individualism and individual rights are ingrained in our collective psyche. And since corporations now have the same rights as individuals, they also have the right to do almost anything they want, including spraying their land with pesticides, building polluting factories on it, or using it as a toxic dump. But although the primacy of individual rights can make it hard to argue successfully for the collective right to a healthy environment, sustained efforts and strong public support can be effective, as shown by the victories of the antislavery, suffrage, and civil rights movements mentioned at the beginning of this section.

Some environmental health and justice groups, such as Earthjustice, the Environmental Health Strategy Center, and Communities for a Better Environment, mention the right to a healthy environment in their mission statements. Others support it implicitly. One group that's been very explicit is the Science and Environmental Health Network (SEHN). SEHN has cleverly extended environmental justice to include the rights of future generations. Building off the Native American belief that collective decisions should take account of the effects on the seventh generation, SEHN states that "People who live today have the sacred right and obligation to protect the commonwealth of the Earth and the common health of people and all our relations for many generations to come."[23] In 2008, SEHN collaborated with the International Human Rights Clinic at Harvard Law School to release a report containing model state constitutional provisions and a model state statute to protect the environmental rights of future generations.[24]

The human rights argument is extremely powerful and could be used by environmental health groups much more often. By arguing that equality and justice should underlie all environmental legislation and policy, the U.S. environmental health movement could strengthen its case considerably. This view seems to be gaining support from several well-known environmentalists including Gus Speth, former White House advisor and retired Yale Law School Dean who, in 2010, said: "Many established environmental issues should be seen as human rights issues—the right to water and sanitation, the right to sustainable development, the right to cultural survival, freedom from climatic disruption and ruin, freedom to live in a non-toxic environment, and the rights of future generations."[25]

INFORMATION ON ENVIRONMENTAL HEALTH INJUSTICE IN THE UNITED STATES

Although environmental justice activists view it primarily as a social critique, they must still make their case using scientific information. Because science is the dominant Western culture's chief way of knowing, the environmental justice movement must use scientific information to demonstrate the reality of disproportional risks and effects, in the same way that the U.S. environmental health movement must use science to demonstrate the reality of environmental health effects. So what's the scientific evidence of environmental injustice?

As discussed in chapter 4, studies conducted in the 1980s and 1990s revealed that environmental injustice is rampant in the United States. Since then, scientists have continued to study the subject, using epidemiological and, more recently, spatially based approaches. Spatially based approaches are particularly powerful because they can visually demonstrate environmental inequities in a way that others can't. Over the years, studies have correlated air pollution,[26, 27, 28, 29] toxic waste dumps,[30, 31] and environmental hazards[32] with income, race, and ethnicity. The Toxics Release Inventory (TRI) is a particularly rich source of information because in the early 1990s, the EPA launched the Risk Screening Environmental Indicators (RSEI) project, which combines TRI data, toxicity information, and data on the size of exposed populations to generate geographically based risk estimates for some chronic diseases. One study using RSEI data showed that the most polluted regions in the United States have higher percentages of African American, Latino, and Asian American residents.[33] Another showed that minority and low-income people face above average environmental health risks and identified the corporations most responsible for emissions to low-income and minority communities.[34] To add to these studies, others indicate that people of color and/or people living in poverty suffer from higher rates of environmentally related diseases, including asthma,[35] cancer,[36] lead poisoning,[37] and cardiovascular disease.[38] For instance, black children have nearly twice the rate of asthma as white children.[39]

The weight of the scientific evidence confirms that environmental injustice is pervasive in the United States. Although it can be difficult to establish unequivocal causal relationships between specific health problems and environmental injustice (see chapter 7), the weight of evidence is overwhelming. The most marginalized groups in society—the poor and people of color—are getting sick and dying from environmental pollutants, and this demands immediate action. To wait for absolute proof of harm would be unethical. Quite rightly, environmental justice activists call for action now, and by working on issues such as toxic sites, the cumulative impacts of urban environmental quality, and climate disruption, they're making their voices heard.

ENVIRONMENTAL JUSTICE ISSUES

Regrettably, the problem of toxic sites hasn't been resolved; it's actually gotten worse. Despite Superfund and other legislation, the problem of toxic sites just keeps getting bigger. Moreover, it's clear that this is a problem for many poor communities and communities of color, not just African American ones. Exacerbated by the decline of the U.S. manufacturing industry, there are thousands of abandoned industrial sites across the country, and many contain toxic chemicals that pose significant threats to health.

The story of the communities and workers affected by ASARCO (the American Smelting and Refining Company) is illustrative. Founded in 1889, ASARCO operated mines, smelters, and refineries in twenty-one U.S. states, as well as in Australia, Mexico, Chile, the Congo, and the Philippines. The company's activities resulted in widespread air, water, and soil pollution, and it was taken to court many times because of this. In fact, ASARCO has been found legally responsible for pollution at about twenty Superfund sites across the United States. One of these sites is at ASARCO's smelter in Ruston, Washington, where residents in a four-county area—about one thousand square miles—have been exposed to arsenic and lead for decades. Many are low income and/or people of color. Despite a massive cleanup effort, many residents are still exposed to contaminated soil. To add to this, ASARCO's company doctor published misleading scientific studies about workers' exposure to arsenic and its health effects.[40] But the events at Ruston were not an isolated incident. In Hayden, Arizona, workers at the ASARCO smelter discovered that the company systematically altered the lung function test results of Mexican-American workers in order to conceal damage to their lungs.[41] Meanwhile, in El Paso, Texas, and East Helena, Montana, workers and community residents were exposed when hazardous waste was secretly incinerated at ASARCO facilities in the 1990s. Although the EPA identified the violations and fined the company, the impacted communities were not informed and nothing was done to monitor the health of workers who handled the waste.[42]

In 2005, ASARCO filed for Chapter 11 bankruptcy, citing environmental liabilities as a primary cause. More than a dozen states and the federal government filed $6 billion in environmental claims against the company,[43] making it the largest environmental bankruptcy in U.S. history. Some four years later, in 2009, ASARCO agreed to pay almost $1.8 billion to ninety communities from twenty-one states for environmental cleanup and monitoring and limited compensation to some of its workers.[44] This figure, however, represents less than 1 percent of the funds originally identified as needed by those affected.[45]

One of the major challenges in Ruston and Hayden was that community residents were afraid to speak out. Even though they were concerned about their health, many people supported the company because they were worried that the smelters would close or they would lose their jobs. When people depend on nearby industries for their income, speaking out can be a very scary thing to do. Legal protection for whistle-blowers is all well and good, but who can blame low income, minority workers for failing to tell the truth about their employers? Terrified of losing their paychecks, they're forced to remain silent about what they know. The result is a vicious cycle in which polluting corporations can continue to damage the health of workers and local communities because the residents who work for them are too frightened to object.

As well as working on toxic sites, environmental justice groups are active on many other issues. While some oppose siting new hazardous facilities; expanding road and rail corridors; air, water, and noise pollution; and hazardous working conditions; others investigate levels of pollutants and clusters of environmentally related diseases. Most of these issues center on the urban environment. In fact, the environmental justice movement is primarily an urban movement, like the environmental health movement. While it's true that migrant farmworker issues, such as pesticide drift, are critically important, most environmental justice work focuses on the urban environment. This is significant because it was precisely these issues that were ignored by the environmental movement of the late 1960s and 1970s (see chapter 3). By joining environmental health activists and laying claim to urban environmental health issues, environmental justice activists filled the vacuum and leveraged public concern into local grassroots activism.

As local environmental health and justice activists worked on specific locally identified issues, the problem of cumulative environmental health effects became very apparent. As discussed in chapter 7, risk assessment doesn't deal with cumulative impacts, so it's not surprising that the environmental justice movement has been critical of risk assessment, just like the environmental health movement. Not only does risk assessment fail to deal with the interactions among multiple environmental hazards, it can't cope with the combined effects of toxic exposures and the socioeconomic stressors experienced by many communities of color and people living in poverty, such as poor quality housing, the shortage of affordable housing, the lack of access to health care, high unemployment rates, low income, limited education and literacy, and the lack of access to healthy food and green spaces. Put all these socioeconomic stressors together with above average exposure to toxic chemicals, and the enormity of the cumulative impacts problem becomes apparent, as does the injustice involved.

In response, several regulatory agencies and/or their advisory boards have acknowledged the importance of cumulative impacts and begun to wrestle with it from a government policy perspective. In 2003, the EPA released a *Framework for Cumulative Risk Assessment*,[46] which was intended to provide a simple structure for conducting and evaluating cumulative risk assessment within the EPA, and in 2004 the EPA's National Environmental Justice Advisory Council offered recommendations on risk reduction in communities with multiple stressors.[47] At the state level, New Jersey's Cumulative Impacts Subcommittee of the Environmental Justice Advisory Council has proposed a series of recommendations to reduce the cumulative impacts of air pollution and identify especially vulnerable communities,[48] and in California, the Office of Environmental Health Hazard Assessment has a Cumulative Impacts and Precautionary Approaches Workgroup to provide advice and there is a *Draft California Communities Environmental Health Screening Tool*, [49] which contains a method to evaluate the cumulative impacts of pollution on communities. Although there's still a long way to go to assess cumulative impacts, these initiatives are a good start. However, rather than waiting for the results of complicated, expensive, and time-consuming cumulative impact assessments, it may be more prudent to simply adopt precautionary approaches.

One environmental justice issue that's rarely considered in cumulative impact assessments is climate disruption. It's already well-established that rising temperatures and increasingly unstable weather patterns disproportionately affect the poor. According to Rajendra Pachauri, chair of the Intergovernmental Panel on Climate Change, "It is the poorest of the poor in the world, and this includes poor people even in prosperous societies, who are going to be the worst hit."[50] U.S. experts have reached the same conclusion about this country. For instance, the U.S. Global Change Research Program has identified people living in poverty as one of the groups most vulnerable to climate disruption,[51] and a University of California study concluded that communities of color and the poor will suffer disproportionately during extreme heat waves, breathe even dirtier air, pay more for basic necessities, and have fewer job opportunities.[52] Given this, it's no wonder that climate disruption has become an important environmental justice issue.

Emerging after the first Climate Justice Summit held in The Hague, Netherlands, in 2000, the climate justice movement has grown rapidly.[53] Leading international coalitions include Climate Justice Now!, established in 2007, and Climate Justice Action, a network that promotes direct action as a tool for local struggles against climate disruption. In the United States, national climate justice organizations include the Environmental Justice Climate Change Initiative, Mobilization for Climate Justice, and the NAACP's Climate Justice Initiative. These three organizations have participants including environmental justice activists, African American and Indigenous organizations, faith-based groups, workers' unions, community groups, and environmental health organizations. Making the case that climate disruption disproportionately affects the poor and people of color, they're helping to build a North American climate justice movement. Their fundamental message is that marginalized communities and populations cannot wait for even more scientific proof of climate disruption; they're already living with its consequences, and they want action now. Arguing that those least responsible are currently experiencing—and will continue to experience—the greatest impacts, they're using a rights-based approach to advocate for social change.

Whether it's working on toxic sites, the cumulative impacts of urban environmental quality, climate disruption, or any other issue, environmental justice activists are critically important in the struggle for environmental health. Their rights-based approach complements efforts to personalize environmental issues (see chapter 6) and adds to scientific arguments for protecting environmental health.

COMMUNITY-BASED RESEARCH

Over the years, many local environmental health and justice groups have had to become their own research experts and learn how to conduct scientific studies in their communities. Dismayed by the way that risk assessment and conventional environmental health science have ignored environmental injustice, they've taken matters into their own hands. Lois Gibbs, the founder of the environmental health movement, was perhaps the first community-based environmental health researcher. In 1978, she investigated disease patterns around Love Canal, and ever since communities affected by pollution have tried to do their own environmental health surveys. Unfortunately, many communities have become frustrated because these types of studies can rarely, if ever, absolutely "prove" that something in the environment has caused a health effect. But community-based environmental health surveys have their uses. Although they can't completely prove causation, they can assess local environmental concerns and collect information about health problems that can be used to identify issues, raise public awareness, and advocate for change.

Many sympathetic environmental health scientists working in universities and research institutes have partnered with communities to study local environmental health problems. For instance, scientists at Columbia University's Children's Environmental Health Center collaborate with West Harlem Environmental Action (WE ACT), and researchers at the University of California have spent years working with migrant farmworkers on pesticides and other environmental health issues. In addition, the federal government is actively supporting partnerships between academic researchers and communities with environmental health problems. For instance, the National Institutes for Environmental Health Sciences provides funding for community-based participatory research, and the EPA is funding community-based research in poor and underserved communities on cumulative impacts.

Many local groups, especially those in "fenceline communities" located next to hazardous industries or facilities, have been analyzing the contaminants in their neighborhoods for years. Because they face greater exposures than other communities, fenceline communities are especially vulnerable and are on the frontlines of environmental health risks. They're the proverbial canaries in the coal mine and can provide an early warning system for the rest of us.

In the mid-1990s, some fenceline communities began to analyze air quality in their neighborhoods, so they didn't need to rely on government agencies or corporations. Calling themselves "bucket brigades," they used a device called a "bucket" to collect samples of air coming across property lines from local industrial facilities. The idea emerged in 1995 after attorney Edward Masry and Erin Brockovich became ill, having been exposed to fumes from a petroleum refinery in northern California. Government authorities claimed their environmental monitors had not detected any problems in the area, leading Masry to hire an environmental engineer to design a low-cost device which community members could use to monitor air quality themselves—the "bucket." This simple and easy-to-use piece of equipment was soon being used by many communities concerned about air pollution from nearby chemical, oil, and gas companies.

Today, there are bucket brigades in at least twenty states, including California, Louisiana, Ohio, Tennessee, Texas, and Virginia, as well as many other countries. One of the reasons that bucket brigades have become so widespread is that they've been supported by an organization called Global

Community Monitor (GCM). GCM provides communities with buckets and other simple, low-cost environmental monitoring technologies and trains residents to use them, so they can understand the effects of pollution on their health and the environment. Over the years, Denny Larson, the founder and executive director of GCM, has worked with numerous bucket brigades and fenceline communities, helping them to monitor polluting industries and negotiate agreements with them.

Community-based research allows ordinary people to study environmental health problems in their own neighborhoods by encouraging them to define the issues, develop and conduct the research, and interpret the results. It builds local capacity, scientific literacy, and empowers communities. In doing so, it democratizes science and puts it into the hands of the public, and this helps to identify environmental health and injustice problems that would otherwise be ignored.

ENVIRONMENTAL JUSTICE STRATEGIES

But identifying environmental health and injustice problems is not enough; environmental justice activists must also demand change, and the single most important strategy they use is collective action. By bringing together local residents to identify their common problems and then mobilizing them to oppose the power-holders, local environmental health and justice activists are first and foremost community organizers. Community organizing is not about external groups coming into a community and telling people what they should be concerned about; it's about local people saying what issues are important to them and then demanding that they be fixed. Collective action is also nonviolent. Based on the ideas of Mahatma Gandhi and Martin Luther King Jr., nonviolent collective action can be a powerful way to achieve social change, as evidenced by the U.S. civil rights movement, the velvet revolution in Czechoslovakia, and the fall of the Soviet Union.

Collective action is, of course, an umbrella term for a huge number of different activities. In his book *The Politics of Nonviolent Action*,[54] social movement theorist Gene Sharp identifies 198 methods of nonviolent collective action. Ranging from marches, protests, and sit-ins to drama, art, and poetry, they're all based on opposing injustice through peaceful persuasion. Over the years, environmental health and justice activists have used most of them in countless communities across the United States. Although there are far too many to mention here, the story of Hilton Kelley,[55] winner of the 2011 Goldman prize,[56] is illustrative.

Hilton Kelley was born and raised in the West Side neighborhood of Port Arthur, Texas. Surrounded by major petrochemical and hazardous waste facilities, this largely African American community has endured hazardous emissions from the Motiva oil refinery, the Valero refinery, the Huntsman Petrochemical plant, the Chevron Phillips plant, the Great Lakes Carbon Corporation's petroleum facility, the Total Petrochemicals USA facility, the Veolia incinerator facility, and the BASF Fina Petrochemicals plant for decades. Hilton Kelley was born and raised there and spent many of his early years living in a public housing project on the fenceline of the Motiva refinery. As an adult, he returned home for a visit and saw the community sickened by industrial pollution, plagued with crime, and teetering on the brink of total economic collapse. He decided to stay and became the leader of the local movement to clean up Port Arthur. Establishing a neighborhood environmental health and justice group, Community In-Power and Development Association (CIDA), he trained local residents to monitor their local air quality using "bucket" technology.

In 2006, when Motiva announced that it would expand its Port Arthur facility into the largest petrochemical refinery in the country, Kelley and CIDA went to work. As a result of their efforts, the company installed state-of-the-art pollution control equipment, agreed to provide health coverage for local residents for three years, and established a $3.5 million fund to help entrepreneurs launch new businesses in the community. He also led a campaign that prevented Veolia Corporation from importing more than twenty thousand tons of toxic PCBs from Mexico for incineration at its Port Arthur plant. Thanks to his leadership, Port Arthur was selected as an EPA environmental justice showcase community.[57] Today, Kelley continues to advocate for stricter environmental regulations on the Texas Gulf Coast and serves on the EPA's National Environmental Justice Advisory Council.

Collective action strategies aren't just about protests and opposing corporate interests head-on, they're also about developing constructive alternatives that address the environmental and economic issues that plague many disadvantaged communities. By working with government agencies and corporations to support community-based economic development, they're encouraging locally and minority-owned companies; training and apprenticeship programs; public investments in community infrastructure, and opportunities for meaningful participation in local decision-making processes. Rather than simply challenging the *status quo*, community-based economic development initiatives are working to create positive, collaborative solutions. The results include a more equitable distribution of wealth, more vibrant sustainable communities, and improved environmental health and justice.

Perhaps the best-known example of this approach is Sustainable South Bronx (SSBx). Established in 2001 by Majora Carter, this environmental justice organization has helped to revitalize the South Bronx area of New York. Among its many achievements are the creation of a riverside park, the installation and maintenance of green roofs by its own for-profit installation company SmartRoofs, and a job training program called the Bronx Environmental Stewardship Program (BEST). SSBx also provides energy retrofitting services to local residents and the FabLab After School program, which seeks to inspire students to become ecologically conscious designers and community problem solvers.

These collective action strategies are being used across the United States. Pointing to the fact that achieving environmental health and justice will require a new sustainable, equitable economy (see chapter 9), as well as new locally based inclusive decision-making processes, they challenge government agencies and corporate interests to rethink unjust and unhealthy economic development models.

THE U.S. ENVIRONMENTAL JUSTICE AND ENVIRONMENTAL HEALTH MOVEMENTS

As noted in chapter 4, it can be difficult to separate the environmental justice and environmental health movements. Although they can be distinguished by the way they frame issues, their origins, and their tactics, they're sometimes seen as a single movement. For instance, Lois Gibbs has called the environmental health movement "a new social justice movement,"[58] and on its web site the Environmental Health Fund, now called Coming Clean, refers to "the health and justice movement."[59] Today, the line between the U.S. environmental health and justice movements can be blurry, just like the line between the environmental movement and the environmental health movement.

The differences between environmental justice and environmental health were more apparent in the 1990s and 2000s when state environmental health organizations stepped up their legislative efforts to reform toxics policies (see chapter 5). As they did this, some local environmental health and justice activists became resentful because they felt that state groups were ignoring already marginalized communities who face greater environmental health risks, discounting the importance of local issues, and scooping up most of the available foundation funding. On the other hand, state groups asserted that stronger legislation and regulations would benefit everyone, including people of color and people living in poverty. Hence, a tension developed between them, at least in the minds of some local activists.

At this time, it also became clear that local groups often lacked the skills and resources they needed. Although early environmental justice groups benefitted from the community organizing skills of civil rights activists, by the 1990s and 2000s many had lost this experience. Moreover, their emphasis on local issues and local action meant that they often worked in isolation from each other and did not have the political clout that comes from building partnerships or participating in coalitions across larger geographic areas. Fully engaged in fighting for their own issues, local environmental health and justice groups often had little time and energy to work with groups located elsewhere. As a result, it was difficult for them to develop or engage with the coalitions being developed by state groups. To add to this, the work of local environmental health and justice groups rarely receives the attention it deserves because it's often overshadowed by larger and better known national and state groups. Getting a new law or policy passed attracts media headlines that community leaders can only dream about. By comparison, organizing for environmental health and justice at the neighborhood level can appear to be unimportant and insignificant. But despite this invisibility problem, local environmental health and justice groups are flourishing in the United States. According to one organizer who has worked at the community level for many years, "The number of local groups is increasing. They continue to spread even though many struggle for survival."[60]

In recent years, the tension between local environmental health and justice groups and larger state and national environmental health organizations has lessened and both are once again appreciating their common goal. This is happening partly because national and state groups are becoming increasingly aware of environmental injustice in the United States and partly because these groups are beginning to understand the importance of locally identified issues and local collective action. Moreover, the groups are beginning to understand the benefits of social change strategies that combine lobbying for legislation on locally identified issues with grassroots collective action on the same issues. The campaigns against fracking and coal-fired power plants (see chapter 5) are good examples of this. By combining lobbying with grassroots collective action, the environmental justice and environmental health movements are working together more closely than ever before. And as the success of their efforts shows, it's a win-win situation.

This chapter has demonstrated the critical role that environmental justice plays in environmental health. Although the U.S. environmental health and justice movements can be distinguished, they share a common goal—environmental health for all—and the line between them is becoming very fuzzy. As local, state, and national groups increasingly work together and coordinate grassroots collective action with lobbying, the resentments of the 1990s and 2000s are fading away to reveal the power of their combined efforts. But even as the U.S. environmental health and justice movements work together more closely, they are challenged by a fundamentally unsustainable and unjust economic system that threatens everyone's environmental health.

NOTES

1. See for example: Robert Bullard. *Unequal protection: Environmental justice and communities of color.* San Francisco: Sierra Club Books (1997); Julian Agyeman, Robert Bullard and Bob Evans (eds). *Just sustainabilities: Development in an unequal world.* Cambridge, MA:

MIT Press (2003); Robert Bullard and Maxine Waters. *The quest for environmental justice: Human rights and the politics of pollution.* San Francisco: Sierra Club Books (2005); Julian Agyeman. *Sustainable communities and the challenge of environmental justice.* New York: New York University Press (2005); Robert Brulle and David Pellow. Environmental justice: Human health and environmental inequalities. *Annual Review of Public Health* 27: 3.1–3.22 (2006); Paul Mohai, David Pellow, and Timmons Roberts. Environmental justice. *Annual Review of Environment and Resources* 34: 405–430 (2009).

2. Julian Agyeman. *Sustainable communities and the challenge of environmental justice.* New York: New York University Press (2005).

3. Susan Cutter. Race, class and environmental justice. *Progress in Geography* 19(1): 111–122.

4. Principles of Environmental Justice. First National People of Color Environmental Leadership Summit held on October 24–27, 1991, in Washington, DC. Available at: http://www.ejnet.org/ej/principles.html. Accessed October 31, 2012.

5. Robert Bullard. Environmental justice in the 21st century, Ch. 1. In: Robert Bullard and Maxine Waters. *The quest for environmental justice: Human rights and the politics of pollution.* San Francisco: Sierra Club Books (2005).

6. EPA Environmental Justice Webpage. Available at: http://www.epa.gov/environmentaljustice/. Accessed October 31, 2012.

7. Julian Agyeman and Robert Evans. Just sustainability: The emerging discourse of environmental justice in Britain? *Geographical Journal* 170(2): 155–164 (2004). p. 155–156.

8. Corpus Iuris Civilis. The Institutes Book 2 of Things, Part 1 Divisions of things. 535 CE, prepared at the order of the Emperor Justinian. Available at: The Medieval Sourcebook http://www.fordham.edu/halsall/basis/535institutes.html. Accessed on October 31, 2012.

9. Carolyn Raffensperger, Burns Weston, David Bollier. Vermont Law School Climate Legacy Initiative Recommendation No. 1. In: Burns Weston and Tracy Bach, *Recalibrating the law of humans with the laws of nature: Climate change, human rights and intergenerational justice.* Vermont Law School and the University of Iowa. (2009). Available at: http://www.vermontlaw.edu/Documents/CLI%20Policy%20Paper/Rec_01%20-%20%28Law_of_Commons%29.pdf. Accessed on October 31, 2012.

10. Declaration of the United Nations Conference on the Human Environment (1972). Available at: http://www.unep.org/Documents.Multilingual/Default.asp?DocumentID=97&ArticleID=1503&l=en. Accessed on October 31, 2012.

11. Fatma Zohra Ksentini, Special Rapporteur. *Human rights and the environment.* United Nations Economic and Social Council, Commission on Human Rights. (1994). Available at: http://www.unhchr.ch/Huridocda/Huridoca.nsf/0/eeab2b6937bccaa18025675c005779c3?Opendocument. Accessed on October 31, 2012.

12. General Assembly adopts resolution recognizing access to clean water, sanitation as human right, by recorded vote of 122 in favour, none against, 41 abstentions. Available at: http://www.un.org/News/Press/docs/2010/ga10967.doc.htm. Accessed October 31, 2012.

13. Article 12 of the Constitution of the Republic of South Africa, as adopted on May 8, 1996, and amended on October 11, 1996, by the Constitutional Assembly.

14. Chapter I, Section 18 of the revised Constitution of Hungary.

15. Constitution of Burkina Faso, adopted on June 2, 1991, as amended on April 11, 2000. Title 1, Chapter 4, Article 29.

16. Carole L. Gallagher. The movement to create an environmental Bill of Rights: From Earth Day, 1970 to the present. *9 Fordham Environmental Law Journal* 107, 120 citing H.R.J. Res. 1321, 90th Congress, 2nd Session. (1968).

17. Carole L. Gallagher. The movement to create an environmental Bill of Rights: From Earth Day, 1970 to the present. *9 Fordham Environmental Law Journal* 107, 120 citing H.R.J. Res. 1205, 91st Congress. (1970).

18. Mary Ellen Cusack. Judicial interpretation of state constitutional rights to a healthful environment. *Boston College Environmental Affairs Law Review* 20(1): 173 (1993).

19. *National Environmental Policy Act* of 1969, as amended. Available at: http://ceq.hss.doe.gov/nepa/regs/nepa/nepaeqia.htm. Accessed on October 31, 2012.

20. Pennsylvania Constitution, Section 27. Available at: http://www.legis.state.pa.us/wu01/vc/visitor_info/creating/constitution.htm. Accessed on October 31, 2012.

21. Hawaii Constitution Article XI, Section 9. Available at: http://hawaii.gov/lrb/con/conart11.html. Accessed on October 31, 2012.

22. Illinois Constitution, Article XI, Section 2. Available at: http://www.ilga.gov/commission/lrb/con11.htm. Accessed on October 31, 2012.

23. Science and Environmental Health Network. *What is future generation guardianship?* Available at: http://www.sehn.org/future.html. Accessed on October 31, 2012.

24. Science and Environmental Health Network and the International Human Rights Clinic at the Harvard Law School, An environmental right for future generations: Model state constitutional provisions & model statute (2008). Available at: http://www.sehn.org/pdf/Model_Provisions_Mod1E7275.pdf. Accessed on October 31, 2012.

25. Gus Speth. Towards a new economy and a new politics. *Solutions* Issue 5 (May 28 2010). Available at: http://thesolutionsjournal.com/node/619. Accessed on October 31, 2012.

26. Manuel Pastor, Rachel Morello-Frosch, and James Sadd. The air is always cleaner on the other side: Race, space, and ambient air toxics exposures in California. *Journal of Urban Affairs* 27: 127–148 (2005).

27. Jayajit Chakraborty. Automobiles, air toxics, and adverse health risks: Environmental inequities in Tampa Bay, Florida. *Annals of the Association of American Geographers* 99: 674–697 (2009).

28. Joshua Fisher, Maggi Kelly, and Jeff Romm. Scales of environmental justice: Combining GIS and spatial analysis for air toxics in West Oakland, California. *Health Place* 12: 701–714 (2006).

29. Marie Lynn Miranda, Sharon E. Edwards, Martha H. Keating, and Christopher J. Paul. Making the environmental justice grade: The relative burden of air pollution in the United States. *International Journal of Environmental Research and Public Health* 8: 1755–1771 (2011).

30. Paul Stretesky and Michael Hogan. Environmental justice: An analysis of superfund sites in Florida. *Social Problems* 45:268–287 (1998).

31. Paul Mohai and Robin Saha. Racial inequality in the distribution of hazardous waste: A national-level reassessment. *Social Problems* 54: 343–370 (2007).

32. Liam Downey. U.S. metropolitan-area variation in environmental inequality outcomes. *Urban Studies* 44: 953–977 (2007).

33. Nicolaas Bouwes, Stephen Hassur, and Marc Shapiro. Information for empowerment: The EPA's risk screening environmental indicators project. Ch. 6 p. 117–134. In: James Boyce and Barry Shelley (eds). *Natural Assets: Democratizing Environmental Ownership.* Washington, DC: Island Press (2003).

34. Martin Ash, James Boyce, Grace Chang, Manuel Pastor, Justin Scoggins, and Jennifer Tran. *Justice in the air.* Amherst, MA: Political Economy Research Group (2009). Available at: http://www.peri.umass.edu/fileadmin/pdf/dpe/ctip/justice_in_the_air.pdf. Accessed on October 31, 2012.

35. Phil Brown, Brian Mayer, Stephen Zavestoski, Theo Luebke, Joshua Mandelbaum, and Sabrina Mccormick. The health politics of asthma: Environmental justice and collective illness experience in the United States. *Social Science and Medicine* 57(3): 453–64 (2003).

36. Elizabeth Ward, Ahmedin Jemal, Vilma Cokkinides, Gopal K. Singh, Cheryll Cardinez, Asma Ghafoor, and Michael Thun. Cancer disparities by race/ethnicity and socioeconomic status. *CA: A Cancer Journal for Clinicians* 54: 78–93 (2004).

37. Michael Kraft and Denise Scheberle. Environmental justice and the allocation of risk: The case of lead and public health. *Policy Studies Journal* 23(1): 113–122 (1995).

38. Robert Brulle and David Pellow. Environmental justice: Human health and environmental inequalities. *Annual Review of Public Health* 27: 103–124 (2006).

39. EPA. *Children's environmental health disparities: Black and African American children and asthma.* Factsheet. Available at: http://yosemite.epa.gov/ochp/ochpweb.nsf/content/HD_AA_Asthma.htm/$File/HD_AA_Asthma.pdf. Accessed October 31, 2012.

40. Marianne Sullivan. Contested science and exposed workers: ASARCO and the occupational standard for inorganic arsenic. *Public Health Reports* 122(4): 541–547 (2007).

41. *Their mines, our stories: ASARCO's environmental practices exposed*. Available at: http://www.theirminesourstories.org/?cat=18. Accessed October 31, 2012.
42. *Their mines, our stories*. Available at: http://www.theirminesourstories.org/. Accessed on March 1, 2012.
43. Asarco begins settlement talks while facing $6 billion in claims. *Mining Engineering* 59(3): 23 (2007).
44. EPA. *Largest environmental bankruptcy in U.S. history will result in payment of $1.79 billion towards environmental cleanup and restoration/Largest recovery of money for hazardous waste clean up ever.* Press Release. December 10, 2009. Available at: http://yosemite.epa.gov/opa/admpress.nsf/d0cf6618525a9efb85257359003fb69d/c40dd49b8eebe5ff85257688006c9c7f!OpenDocument. Accessed October 31, 2012.
45. Mara Kardas-Nelson, Lin Nelson, and Anne Fischel. Bankruptcy as corporate makeover: ASARCO demonstrates how to evade corporate responsibility. *Dollars and Sense* May/June, 2010. Available at: http://www.dollarsandsense.org/archives/2010/0510kardas-nelson-nelson-fischel.html. Accessed October 31, 2012.
46. EPA. *Framework for cumulative risk assessment*. EPA/630/P-02/001F (May 2003). Available at: http://www.epa.gov/raf/publications/pdfs/frmwrk_cum_risk_assmnt.pdf. Accessed October 31, 2012.
47. National Environmental Justice Advisory Council. *Ensuring risk reduction in communities with multiple stressors*. U.S. EPA (2004). Available at: http://www.epa.gov/environmentaljustice/resources/publications/nejac/nejac-cum-risk-rpt-122104.pdf. Accessed on October 31, 2012.
48. Cumulative Impacts Subcommittee of the Environmental Justice Advisory Council to the New Jersey Department of Environmental Protection. *Strategies for addressing cumulative impacts in environmental justice communities*. (2009). Available at: http://www.state.nj.us/dep/ej/docs/ejac_impacts_report200903.pdf. Accessed on October 31, 2012.
49. California Environmental Protection Agency, Office of Environmental Health Hazard Assessment. *Draft California communities environmental health screening tool (CalEnviroScreen): Proposed methods and indicators*. (July 30, 2012). Available at: http://oehha.ca.gov/ej/pdf/DraftCalEnviroScreen073012.pdf. Accessed October 31, 2012.
50. As quoted in: *UN climate change impact report: Poor will suffer most* (April 6, 2007). Available at: http://www.ens-newswire.com/ens/apr2007/2007-04-06-01.asp. Accessed October 31, 2012.
51. United States Global Change Research Program. *Global climate change impacts in the United States*.(2009). Available at: http://www.globalchange.gov/publications/reports/scientific-assessments/us-impacts. Accessed October 31, 2012.
52. Rachel Morello-Frosch, Manuel Pastor, James Sadd, and Seth Shonkoff. *The climate gap: Inequalities in how climate change hurts Americans & how to close the gap*. (2009). Available at: http://dornsife.usc.edu/pere/documents/The_Climate_Gap_Full_Report_FINAL.pdf. Accessed October 31, 2012.
53. Jethro Pettit. Climate justice: A new social movement for atmospheric rights. *IDS Bulletin* 35(3): 102–106 (July 2004).
54. Gene Sharp. *The politics of nonviolent action, Part two: The methods of nonviolent action*. Boston: Porter Sargent (1973).
55. Steve Lerner. *Port Arthur, Texas: Public housing residents breathe contaminated air from nearby refineries and chemical plants* . The Collaborative on Health and the Environment. Available at: http://www.healthandenvironment.org/articles/homepage/1008. Accessed October 31, 2012.
56. The Goldman Environmental Prize. Prize Recipient Hilton Kelley, 2011 North America. Available at: http://www.goldmanprize.org/2011/northamerica. Accessed October 31, 2012.
57. EPA Environmental Justice. *Region 6 EJ showcase community: Port Arthur, TX*. Available at: http://www.epa.gov/compliance/ej/grants/ej-showcase-r06.html. Accessed October 31, 2012.
58. Lois Gibbs. Citizen activism for environmental health: The growth of a powerful new grassroots health movement. *The Annals of the American Academy of Political and Social Science* 584: 97–109 (2002).

59. Environmental Health Fund. Background. Available at: http://environmentalhealth-fund.org/about.background.php. Accessed October 31, 2012.
60. Denny Larson. Personal communication. May 26 2010.

Chapter Nine

Changing Economics, the Markets, and Business

Although anthropocentrism is an underlying cause of many environmental health problems, Western culture's economic system is a close second. As discussed in the Introduction, economic development threatens the planet's life support systems and human health. The unsustainable exploitation of the earth's natural resources combined with the generation of unprecedented amounts of pollution and waste are a recipe for disaster. It's clear that we need a new type of economy—one that is sustainable and equitable. We need an economy that will meet everyone's basic human needs, be fair and just, respect ecological limits, provide for the needs of future generations, and respond to local, as well as global priorities. This type of economy would also protect and enhance environmental health.

This vision is no pipe dream. Even though over half of the world's largest financial systems are profit-hungry corporations,[1] the early signs of a new economic system are everywhere. For instance, there's a huge increase in the numbers of farmers' markets and community-supported agriculture in the United States, local trading and barter systems are springing up in many communities, peer-to-peer transactions are cutting out corporate middle men, benefit and low-profit limited liability corporations (also known as B-corps and L3Cs) are working to redefine business in terms of social and environmental performance, and there's growing interest in cooperatives and alternative forms of business ownership.

The emergence of a new economy presents an opportunity for the U.S. environmental health movement. Even though it could take many decades of dedicated work to transform free market economics, and success is far from guaranteed, a new sustainable, equitable economy could protect and improve environmental health. This suggests that the U.S. environmental health movement should be an active participant in work to change economics, the markets, and business. However, it has been rather conflicted about whether and how to engage. Some believe that a combination of business tools and market mechanisms will be sufficient to achieve environmental health, while others blame capitalism for almost everything and would like to replace the entire system. This tension can be seen in the movement's relationship with the "green economy." Although many agree that green jobs are the solution to high unemployment and deteriorating environmental conditions, others question the concept, claiming that it fails to address the real issue that continuous economic growth is impossible on a finite planet. Some try to ignore economics altogether, but this is a big mistake because it is the dominant ideological framework of our time, permeating all aspects of Western culture. From social values to personal actions, economics shapes our individual and collective lives. Trying to ignore economics is like trying to ignore the proverbial elephant in the middle of the room; the longer one tries to pretend it's not there, the more problematic it becomes.

In recent years, the U.S. environmental health movement has started to work much more intently on changing economics, the markets, and business. Using a variety of methods, such as drawing attention to the costs of environmentally related diseases and disabilities, launching market campaigns that target economic sectors using toxic chemicals, advocating for green chemistry and safer materials, promoting socially responsible investing, and developing partnerships with business, it's now promoting economic, as well as social change. This chapter looks at each of these methods and how they are being used.

THE COSTS OF ENVIRONMENTALLY RELATED ILLNESS

It has been said that economics is the only subject in which two people can win a Nobel Prize for apparently contradictory research. But all joking aside, economics is a deadly serious business especially when it comes to environmental health. Although economic development has contributed to improvements in health over the past century, these gains are not sustainable over the long term because they've been purchased at the expense of the environment.

This situation has arisen because conventional economics fails to take account of the environmental and health costs of doing business. As well as failing to include the cost of the goods and services provided "free" by nature,[2] conventional economics ignores the costs of diseases and disabilities caused by poor environmental quality. These costs are not considered important or even relevant. Corporations simply disregard them. Focusing exclusively on narrowly defined income and expense categories, their balance sheets don't take account of the costs of treating the asthma, cancer, birth defects, learning disabilities, and other health problems caused by their products and activities. They only consider their own bottom lines. This is because the economic system is structured so that corporations are only accountable to their shareholders, not to ordinary people and the environments in which they live. But it isn't only corporations that ignore the environmental and health costs of doing business. Governments do exactly the same thing. They rarely acknowledge the costs of environmentally related diseases in their cost-benefit analyses, even though it's ordinary citizens who end up paying the price through their taxes and health insurance premiums.

On the other hand, the costs of taking action to protect environmental health often weigh heavily in corporate and government decision making. Corporations argue they can't afford to install pollution control equipment or use safer manufacturing processes because it's too expensive. But the truth is that protecting environmental health only appears expensive because the costs of inaction aren't considered. Cost-benefit analyses should always take account of both sides of the metaphorical coin—the costs of taking action to protect environmental health and the costs of the illness and disease resulting from maintaining the *status quo*. Otherwise, decisions are based on only partial information. This is why some U.S. environmental health groups are now drawing attention to the ignored side of the coin—the economic costs of environmentally related diseases and disabilities.

Although including these costs in government and corporate decisions may appear to be a radical and impractical idea, it's not new. About 160 years ago, Charles Dickens argued that the high cost of typhus in London—£440,000 in 1848 alone—should be considered in decisions about whether to implement new public health measures. He commented: "This cold-blooded way of putting the really appalling state of the case is, alas, the only successful mode of appealing. . . . His heart is only reached by his pocket."[3] Little has changed since then. Placing an economic value on people's suffering may be "cold-blooded," but it's as necessary now as it was in Dickens's time. Today, as in Victorian England, money is almost always the single most important factor in corporate and government decision making. So, what is the price tag for environmentally related diseases and disabilities?

Until recently, this has been a difficult question to answer for two reasons. First, environmental health scientists haven't been able to estimate how much disease is caused by poor environmental quality, and second, the economic costs of different health conditions have been difficult to quantify. But over the past twenty-five years, the situation has changed: there's now a growing scientific consensus on the proportions of different diseases and disabilities that are associated with environmental quality, and in 2006, the World Health Organization published a major report on the subject.[4] As well, economists working for the health insurance industry have developed "cost of illness" models for many different diseases that include the direct health-care costs, such as hospitalization, physician and nursing services, and prescription medications, as well as the indirect costs, such as lost productivity, premature death, and needs for social services and special education. These two developments have paved the way for studies on the economic costs of environmentally related diseases and disabilities.

Studies on the economic costs of environmentally related diseases and disabilities generally fall into three categories. The first type of study focuses on the economic benefits of reducing exposure to individual toxic substances, especially lead and methyl mercury. Because lead reduces children's IQ, decreasing childhood exposure will eventually result in increased adult earnings potential[5, 6] and worker productivity.[7] One study found that each dollar spent on programs to reduce children's exposure to lead paint results in a return of $17 to $221, equivalent to net national savings of $181 to $269 billion.[8] Revealing that exposure to this single substance costs the economy billions of dollars each year, these studies helped make the case against lead. Other studies have examined the economic costs of environmentally related diseases and disabilities attributable to exposure to methyl mercury emissions from coal-fired power plants[9] and the health costs of mercury-contaminated food sources in the Arctic.[10]

A second type of study looks at the costs of air pollution. Early research in Allegheny County, Pennsylvania, estimated the cost of hospitalization due to poor air quality at almost $10 million in 1972 dollars.[11] Since then, numerous studies have examined the costs of air pollution. For instance, one study estimated that the pollution caused by coal burning power plants cost $53 billion nationally,[12] and a California State University study showed that the health effects of air pollution cost the California economy more than $28 billion annually.[13] Other studies have looked at the economic benefits of reducing air pollution and compared them with the costs of implementing new air quality regulations. For instance, a 2011 study found that the United States could save $15 million annually in reduced health-care costs from hospitalizations of children with bronchiolitis living in urban areas if levels of fine particulate matter were reduced by 7 percent below the standard,[14] and another found that proposed or recently adopted air quality regulations would cost $195 billion, but would bring over $1 trillion in health and environmental benefits, particularly for minorities and the poor.[15] This is a huge return on investment. Related to these air pollution studies, research on the health costs of climate disruption is now underway, with one of the first studies estimating the health costs of six climate disruption-related events in the United States at $14 billion.[16]

A third type of study examines the costs of environmentally related diseases and disabilities in children. Published in 2002, the first major national study of this type estimated the environmentally attributable costs of just four childhood diseases and disabilities—asthma, cancer, neurobehavioral disorders, and lead poisoning—at $54.9 billion a year in 1997 dollars.[17] This research was followed by similar studies in several states including Massachusetts,[18] Washington,[19] and Oregon.[20] In 2011, the national costs of environmentally related diseases in children were updated to $76.6 billion.[21]

These studies reveal that the costs of environmentally related diseases and disabilities are staggering, represent a huge drain on the U.S. economy, and are much less than the costs of implementing new regulations to improve environmental quality. The evidence is overwhelming. Even if the estimates are not completely accurate, they offer compelling proof that reducing pollution would save billions of dollars a year and significantly improve environmental health.

The U.S. environmental health movement is now using these and other similar studies to make an economic case for protecting environmental health. In fact, even the EPA is using this type of information in its educational materials.[22] Complementing the movement's other strategies, this relatively new line of argument is important because it gets the attention of legislators. By showing them that regulations that reduce pollution are extremely cost effective, economic studies can be powerful persuaders, especially in these days of shrinking government budgets and concerns about the cost of health care.

But even as the movement uses estimates of the financial costs of environmentally related disease and disability, it's important to remember monetary valuations are not the whole story. Indeed, they're only a small piece of it because they can never take account of the psychological and emotional costs of disease to the victims, or to their families, friends, and communities. Economic studies can be persuasive, but they can only measure health costs in dollars and cents; they can never capture the costs of the human suffering. This suggests that although these studies can help to convince legislators to take action, they must always be supplemented with other approaches.

MARKET CAMPAIGNS: OVERVIEW

Another approach being used by the U.S. environmental health movement is market campaigns that target the corporations and people who use toxic chemicals. Realizing that they could achieve social change by persuading downstream users and consumers to stop using products containing toxic chemicals, movement leaders decided to focus on several large economic sectors, including green builders (see chapter 5), personal care products (see chapter 6), PVC products and packaging, electronics, the health sector, and schools. This strategy has been very successful because it complements the movement's head-on attacks on the chemical industry. By reducing the demand for toxic chemicals, it stimulates the development of safer chemicals, thereby changing the entire chemical industry.

Market campaigns are a more sophisticated version of the consumer boycotts used by many previous social movements. The first U.S. consumer boycott targeted the abysmal working conditions in the clothing industry at the end of the nineteenth century. Led by Florence Kelley (1859–1932), a member of the Settlement movement, resident of Hull House (see chapter 2), and general secretary of the National Consumers' League (NCL), it persuaded consumers to stop buying clothing from businesses that did not display the NCL's White Label, which was only given to stores that purchased from manufacturers who provided a satisfactory working environment. As well, members of the NCL were authorized as factory inspectors to ensure compliance with federal workplace legislation. At the height of its success, in 1904, the NCL licensed sixty factories.[23] Ever since then, boycotts have been a very popular strategy for changing corporate behavior. Often targeting individual companies or products, they have a good track record of success. Other U.S. consumer boycotts include campaigns against McDonalds, BP, Shell Oil, and Nike. One of the best publicized was the Grapes of Wrath campaign, launched by Cesar Chavez in 1984 (see chapter 4).

Although U.S. boycotts can influence U.S. consumers, international boycotts are even more persuasive. Because there are so many transnational corporations, they can be an important global strategy for social change. The first international consumer campaign on environmental health was the one against Nestlé, the Swiss baby food manufacturer. Launched in the United States in 1977 and quickly spreading to Europe, it was prompted by concerns that the company was aggressively marketing its infant formula in developing countries and that this was harming children's health.[24] Campaign organizers, including the International Baby Food Action Network, argued that babies were being fed weak solutions of formula and so were not receiving adequate nutrition and that the formula was being made up with contaminated water, resulting in infant illness and death. They also asserted that even if formula was prepared properly, it didn't contain many of the beneficial substances naturally present in breast milk, such as antibodies. At first, the Nestlé campaign was very successful. By 1980, the company's profits had declined by 16 percent,[25] and in 1981 the World Health Assembly adopted the *International Code of Marketing Breast Milk Substitutes*,[26] which Nestlé agreed to abide by. Shortly after this, the boycott was officially suspended. But in 1989, organizers reinstituted it because the company's marketing practices were still unethical. Today, the boycott remains in effect, and in 2011, it was relaunched in the Asia-Pacific region with the support of nineteen international nongovernmental organizations, including World Vision, Save the Children, and Oxfam.

The U.S. environmental health movement has extended the idea of consumer boycotts in several ways. First, as well as advocating outright boycotts, it is encouraging consumers to convince corporations to make their products safer. In contrast to the oppositional stance implicit in traditional boycotts, this approach is much more positive. By using consumer activism to persuade corporations to change their ways, it's a constructive approach to changing the markets. This can be seen in the movement's work on electronics.

MARKET CAMPAIGNS: ELECTRONICS

The electronics industry is huge and uses many different toxic chemicals to manufacture its ever increasing panoply of products. Whether they are considered to be electronic equipment, such as smartphones and iPods, or whether they are components of other types of equipment, such as automobiles and washing machines, electronics dominate our lives. In the United States alone, the Consumer Electronics Association estimates that shipments will total about $216 billion in 2013.[27] Manufacturing these products uses somewhere between five hundred and one thousand different substances, many of them toxic. Containing chlorinated solvents, brominated flame retardants, PVC, lead, mercury, cadmium, beryllium, chromium, phthalates, and other toxic chemicals, electronic products and components are a huge problem.

Given the electronics industry's size and reliance on toxic chemicals, it was a natural choice for a market campaign, and who better to focus on than Apple, one of the world's largest electronics companies. So in 2006, Greenpeace, already an expert in consumer boycotts, launched its *Green My Apple* campaign. This ingenious campaign used the Apple's infamously loyal customer base to persuade the company to phase out the worst toxics from its products and to improve its recycling practices. According to the Greenpeace web site, "we decided this was to be a very different Greenpeace campaign, one in which we would turn over the reigns to Apple's customers. We would stand in the shoes of Apple fans, we would speak as fellow believers in the wizards of Cupertino,[28] and we'd try to channel waves of 'Apple Love' at corporate headquarters."[29] The campaign opened with the words "We love Apple,"[30] which were featured on the *Green My Apple* web site. To encourage the company to become greener, Greenpeace gave away some of its logos and designs so that people could create their own campaign materials. Apple users responded enthusiastically by creating videos, images, blogs, and merchandise items—all encouraging Apple to take action. The campaign worked. In May 2007, only eight months after its launch, Steve Jobs, cofounder and CEO of Apple, announced that the company would phase out brominated fire retardants (BFRs) and polyvinyl chloride (PVC) and create a system for U.S. Apple users to return their products for safe recycling.[31] To add to this, Greenpeace publishes a *Guide to Greener Electronics* which ranks fifteen electronics manufacturers on their policies and practices to reduce their carbon footprint, produce less toxic products, make their operations more sustainable, and offer recycling and takeback programs.

Advocating for recycling and takeback programs is a key strategy in the U.S. environmental health movement's market campaign on electronics. This makes sense because electronic waste, also known as e-waste, is the most rapidly growing component of the municipal solid waste stream. In 2006, the United Nations Environment Programme estimated that the world produced between twenty and fifty million metric tons of e-waste a year,[32] and in 2009, the EPA estimated that Americans threw away about thirty million computers, twenty-three million TVs, and 130 million mobile devices,[33] and that's not including all the outdated products in storage that will probably never be used again. Not surprisingly, the United States generates more e-waste than any other country in the world, an estimated three million metric tons a year,[34] and a huge amount is exported, often illegally, to Asia and Africa. There, workers, some of whom are children, are exposed to a cocktail of toxic chemicals as they disassemble or burn the e-waste in primitive attempts at recycling.

In the 1990s, environmental health groups began lobbying for e-waste takeback and recycling programs and a ban on exports. In the United States, the Electronics TakeBack Coalition has led the charge. Comprising many environmental health groups, including the Center for Environmental Health; the Center for Health, Environment and Justice (CHEJ); and the Breast Cancer Fund, this national coalition advocates for extended producer responsibility, so that electronics manufacturers and brand owners are required to take responsibility for the entire lifecycle of their products. Focusing on the state level of government, the Electronics TakeBack Coalition has helped to pass legislation requiring e-waste recycling in twenty-five states, covering 65 percent of the U.S. population. At the international level, groups such as the Basel Action Network and Greenpeace have been working hard to ban exports of e-waste, and in 2011, the *Responsible Electronics Recycling Act* was introduced into Congress.[35, 36]

The U.S. environmental health movement is also lobbying for extended producer responsibility and takeback programs for other types of products, such as pharmaceuticals, paints, beverage containers, tires, carpets, and mercury-containing items. Based on the idea that manufacturers should be fully accountable for their products, these types of programs help to make the transition from a throwaway society to one that has zero waste. Furthermore, because extended producer responsibility and takeback programs are usually financed and managed by manufacturers themselves, they make it financially advantageous for manufacturers to make less harmful products that can be easily recycled or reused, like the Smart Car, which is designed for easy disassembly and is 95 percent recyclable. By encouraging extended producer responsibility and takeback programs, the U.S. environmental health movement is encouraging sustainable design, helping to reduce toxic wastes, and changing economics, the markets, and business.

MARKET CAMPAIGNS: PVC PRODUCTS AND PACKAGING

At about the same time as Greenpeace's *Green My Apple* campaign was in full swing, the Center for Health, Environment and Justice (CHEJ) and other U.S. environmental health groups launched a national market campaign against the use of PVC in products and packaging. PVC is used in countless products and a lot of packaging. In fact, the annual consumption of PVC resin in North America is a whopping 6.4 million metric tons, or almost 14.2 billion pounds.[37] From an environmental health perspective, it's one of the most hazardous plastic around, posing risks during its manufacturing, use, and disposal. For consumers, PVC is almost impossible to avoid because it's widely used in the construction, automotive, health care, and electronics industries, as well as in consumer products, such as furniture, clothes, credit cards, shower curtains, children's toys, food containers, and packaging. Although packaging is one of the smaller uses of PVC, it's one of the most significant because PVC packaging is used to wrap a wide variety of consumer products, including meats, vegetables, and other foods.

But it's the risks from PVC's manufacturing and disposal that are particularly alarming. The production of PVC requires several toxic chemicals, including chlorine, carcinogenic vinyl chloride monomer (VCM), and ethylene dichloride (EDC). The process also uses toxic additives such as phthalates, lead, cadmium, and tin, and it releases dioxins. These chemicals can cause numerous health effects including cancer, neurodevelopmental effects, and reproductive effects. And then there's disposing of the stuff. Land-based disposal of PVC can result in groundwater pollution, and incineration is a major source of dioxins to the atmosphere.

Given these risks, launching a market campaign on PVC packaging and products made good sense. So in March 2006, a coalition of sixty-six health and environmental organizations, led by CHEJ, sent letters to Target and other big-box retailers asking them to phase out PVC from their products and packaging. Wal-Mart and Sears/K-Mart agreed, but Target did not. As a result, Target was targeted. As the fifth largest retailer in the United States, with over $59 billion in annual revenues and about fifteen hundred stores, Target was an obvious choice. In October 2006, CHEJ organized a "day of action" at more than thirty Target stores across the United States. The day featured protests, banners, activists wearing hazmat suits, a twenty-five-foot inflatable rubber ducky, and the release of an animated video called *Sam Suds*.[38] The video was viewed online more than forty thousand times within its first month, and it was screened at a national conference to an audience of over eleven thousand. In the following months, CHEJ collaborated with Working Assets, a progressive telephone company, to send out an action alert to over 250,000 subscribers. They deluged Target with tens of thousands of letters, phone calls, and petitions—all calling for the company to take action on PVC. In the spring of 2007, a second even larger national day of action was scheduled to coincide with the company's annual shareholder meeting. Shortly after in November 2007, Target said it would reduce the use of PVC in many of its items, including infant products, children's toys, shower curtains, packaging, and fashion accessories.[39]

As a result of this work, CHEJ and its partners have shifted an entire market sector away from PVC. This substance is now being reduced and phased out by many major companies, including Adidas, Aveda, Body Shop, Boots, Brio, Bristol-Myers, Costco, Crabtree & Evelyn, Dell, Evenflo, Estée Lauder, Evian, Firestone Building Products, Gerber, H&M, Hewlett Packard, Honda, Ikea, Johnson & Johnson, Kaiser Permanente, Kiss My Face, Lamaze Infant Development, Lego Systems, Limited Brands (Bath & Body Works, Victoria's Secret), Marks & Spencer, Microsoft, Nike, Nokia, Puma, Samsung, SC Johnson, Sharp, Toyota, and Volvo.

MARKET CAMPAIGNS: THE HEALTH-CARE SECTOR

A third market campaign is the U.S. environmental health movement's work on the health sector. With national health-care expenditures expected to reach almost $4.8 trillion a year by 2021[40]—about one-fifth of the country's gross domestic product—it's one of the largest and most influential sectors of the U.S. economy. But not only is the health-care sector very large and powerful, it's also a major user of toxic chemicals. In fact, it's all but impossible to avoid toxic chemicals in health care. Present in IV tubing and other plastic equipment, medical devices, disinfectants, cleaners, pesticides, and electronics, it's hard for patients and their caregivers to avoid exposure to carcinogens, endocrine disruptors, developmental toxins, and skin and eye irritants. And when these products are thrown away, the hazardous substances they contain end up in the environment. Given all this, there's no doubt that the health-care sector is an important source of environmental health hazards.

Several health-care coalitions are attempting to reduce these hazards. Key among them is Health Care Without Harm (HCWH). Established in 1996 to advocate for the closure of medical waste incinerators (see chapter 4), it has grown into a broadly based international coalition of almost five hundred organizations in more than fifty countries. Today, the members of HCWH include hospitals and health-care systems, medical professionals, community groups, patients, labor unions, environmental and environmental health organizations, and faith groups. What unites them is a commitment to a health-care sector that does no harm. Consistent with the Hippocratic oath, HCWH members are in unanimous agreement that the health-care sector should not cause any harmful environmental health effects. Not only has HCWH helped to close thousands of hospital waste incinerators in the United States, it's been very successful in other ways too. For example, HCWH played a leading role in virtually eliminating the market for mercury-based medical equipment in the United States and banning mercury thermometers in the European Union. It also helped to create new markets for alternatives to PVC and diethyl hexyl phthalate. Currently, HCWH is working on reducing the use of toxics in health-care products, on climate disruption, green building, energy, healthy food systems, and other issues.

Another health-care coalition working to improve environmental health is Practice Greenhealth. Created in 2008, when an organization called Hospitals for a Healthy Environment (H2E) was restructured, it's now working with U.S. health-care facilities and corporations that want to become more sustainable. Practice Greenhealth has over a thousand members, including hospitals and health-care systems; health-care providers; manufacturers and service providers; architectural, engineering, and design firms; group purchasing organizations; and affiliated nonprofit organizations. Offering tools, resources, training programs, consulting services, and awards on a variety of issues, it's an educational and networking organization. Together, Health Care Without Harm and Practice Greenhealth are raising awareness and working to promote safer and less toxic products. They are both extremely valuable allies for the U.S. environmental health movement.

GREEN CHEMISTRY AND SAFER MATERIALS

Much of the U.S. environmental health movement's work on market campaigns is based on promoting safer, less toxic products. After the 2004 *Louisville Charter* (see chapter 5), it stepped up work on the subject and became a strong supporter of green chemistry and safer materials. The Charter's call for "altering production processes, substituting safer chemicals, redesigning products and systems, rewarding innovation and re-examining product function" had a powerful effect, and many groups began to advocate for alternatives. Growing out of earlier work on toxics use reduction, pollution prevention (see chapter 4), and precaution (see chapter 7), green chemistry offers a positive solution.

One of the leading proponents of green chemistry is Paul Anastas, who first coined the term in 1991. He went on to establish the Presidential Green Chemistry Challenge Awards and the Green Chemistry Institute, which is now part of the American Chemical Society. In 1998, Anastas paired up with chemist John Warner to write *Green Chemistry: Theory and Practice*,[41] the first major book on the subject, which defined green chemistry as "the utilization of a set of principles that reduces or eliminates the use or generation of hazardous substances in the design, manufacture and application of chemical products." The twelve principles articulated by Anastas and Warner include preventing waste, designing low- or nontoxic chemicals and products, and using renewable feedstocks. Terry Collins is another leading green chemistry researcher. Internationally recognized for his work, he developed a new type of catalyst to replace the chlorine-based oxidant catalysts that contribute to the formation of dioxins and other toxic chlorinated substances. In 1992, he taught the first course on green chemistry at Carnegie Mellon University, where he is director of the Institute for Green Science.

By the early 2000s, green chemistry was taking the academic world by storm. Seen as the next wave in chemical engineering, it's now being taught in universities in Australia, Brazil, Canada, China, India, Italy, Japan, the UK, and the United States. In the United States, dozens of university and college courses offer green chemistry lectures, readings, assignments, and lab exercises, and a few offer degrees in green chemistry. But green chemistry isn't only blackboard conjecture; it's a multibillion-dollar business that's growing fast. According to one industry report, green chemistry is expected to increase from a $2.8 billion business in 2011 to approximately $100 billion by 2020,[42] and it's already been applied in more than twenty economic sectors—everything from aerospace to agriculture and from pharmaceuticals to plastics. Although corporate greenwashing[43] is widespread, an increasing number of businesses are trying to do the right thing and use green chemistry and safer materials, and the U.S. environmental health movement is happy to help.

A few U.S. environmental health groups are working directly with companies and industrial sectors to encourage green chemistry. One of them is Clean Production Action (CPA). With offices in Massachusetts, New York, and Quebec, CPA is translating a "systems-based vision of clean production into the tools and strategies NGOs, governments and businesses need to advance green chemicals, sustainable materials and environmentally preferable products."[44] It does this by offering assessment and evaluation tools, case studies, research, educational materials, and training services. It also coordinates the Business-NGO Working Group for Safer Chemicals and Healthy Materials (see the section of Partnerships with Business) and advocates for legislation on toxics and green chemistry. Other environmental health groups working to promote green chemistry include the Great Lakes Green Chemistry Network and GreenBlue.

There's a lot of work to do on green chemistry, especially at the federal level. Current legislation could be holding back new safer products because, as discussed in chapter 3, the *Toxic Substances Control Act* (1976) contains different requirements for existing and new chemicals. Although there aren't any testing requirements for existing substances (substances that were on the market before 1976), there are requirements for new ones. These differential provisions have the effect of discouraging green chemistry, despite the fact that it would benefit environmental health. It would make more sense to level the playing field so that the manufacturers of all substances—new and existing ones—are required to prove a common standard of safetylike the REACH system in Europe (see chapter 5). In addition to coping with an unfair regulatory regime, companies working on green chemistry struggle to find the funds they need for research and development (R&D). Unlike nanotechnology, which has received billions of dollars in federal funding, R&D on green chemistry struggles to get by on nickels and cents. At the state level, the situation is a little better, and at least four states, including California, Connecticut, Michigan, and Minnesota, have adopted policies supporting green chemistry, while others have held conferences, support research, or offer guidance on the subject.

The U.S. environmental health movement's efforts to promote green chemistry and safer materials mark an exciting trend. Not only do they amplify the potential of precautionary approaches (see chapter 7), they also offer positive alternatives to toxic chemicals and polluting industrial processes. For too many years, the movement has been accused of being negative and failing to propose positive alternatives. Despite work to promote recycling and renewable energy as alternatives to waste incineration and fossil fuels, it has been perceived as obstructing progress. Now, the U.S. environmental health movement's support for green chemistry and safer materials is helping to put an end to this view.

The next step may be to strengthen the business case for green chemistry and safer materials. By making a stronger economic argument, the movement could put forward an even more persuasive rationale. It's not enough to promote the environmental health benefits of green chemistry and safer materials; it's also necessary to promote their economic benefits, and there are plenty of them. By reducing resource costs, process inefficiencies, and waste disposal costs, green chemistry can improve corporate bottom lines. To add to this, it can help companies to increase their market share by appealing to green consumers by enhancing corporate credibility. By making a strong economic case for green chemistry and safer materials, the U.S. environmental health movement could influence the chemical manufacturing industry even more.

SOCIALLY RESPONSIBLE INVESTING

As well as working to change the markets for products and promoting green chemistry and safer materials, some U.S. environmental health groups are working to change financial markets through socially responsible investing (SRI), also known as sustainable responsible impact. Based on investing in companies that help to protect the environment, health, and human rights and avoiding those that do the opposite, it seeks to maximize social as well as financial returns.

Today, SRI is booming, despite the economic recession. According to the Forum for Sustainable and Responsible Investment, at the end of 2011 $3.74 trillion in U.S.-domiciled assets were being managed using SRI tools, equivalent to more than one out of every nine dollars under professional management in the United States.[45] This is a 22 percent increase since 2009 and a 486 percent increase since 1995. Although the SRI market is dominated by large institutional investors, there are many individual investors who rely on SRI mutual funds, such as Calvert, Domini, Parnassus, and Pax World. In 1995, there were only 55 SRI funds in the United States with $12 billion in assets, but by 2012, this had grown to 750 funds with assets of over $1 trillion.[46]

Like consumer boycotts, SRI has been used as a strategy to promote social change for a very long time. In 1758, the Philadelphia Yearly Meeting of the Religious Society of Friends (Quakers) forbade its members from investing in or profiting from the slave trade.[47] Not much later, John Wesley (1703–1791), the founder of Methodism, outlined the two basic principles of SRI—to ensure that one's own business practices are not harmful, and to avoid businesses with harmful practices.[48] Religious organizations have been in the forefront of SRI ever since, although they've now been joined by trade unions, civil rights activists, environmentalists, and others who want to make corporations more accountable for their actions.

Socially responsible investing has been used as a tool to protect environmental health since the 1990s. Soon after the 1989 Exxon Valdez disaster, in which a massive oil spill destroyed the ecosystem of Prince William Sound in Alaska, a group of investors, environmental organizations, and other public interest groups came together to create the Valdez Principles[49] (later known as the CERES Principles). Designed to be an environmental code of conduct for corporations, the CERES Principles include the need to protect environmental health, stating: "We will strive to minimize the environmental, health and safety risks to our employees and the communities in which we operate" and "We will reduce and where possible eliminate the use, manufacture or sale of products and services that cause environmental damage or health or safety hazards." In the 1990s, many corporations endorsed the CERES Principles, but they have now been superseded by the International Standards Organization's (ISO) 14000 series of environmental management standards.

Over the years, financial managers involved in SRI have developed a tool box, comprising these main tools:

- Screening individual corporations, financial portfolios, and mutual funds to ensure they meet predetermined environmental, health, and social criteria;
- Shareholder engagement in which shareholders file resolutions on environmental, health, and social issues at corporate annual meetings; and
- Community investing which supports local economies by directing investment dollars to local organizations that are traditionally underserved by banks and other financial institutions.

With the increasing popularity of SRI and growing concern about environmental health, it's no wonder that there's an organization dedicated exclusively to using SRI to benefit environmental health. Established in 2004, the Investor Environmental Health Network (IEHN) comprises the investment managers of over a dozen socially responsible funds, who are advised by a panel of leading environmental health groups. By early 2008, IEHN members managed more than $41 billion in assets, and in 2009, IEHN was instrumental in persuading McDonalds, the largest buyer of potatoes in the United States, to promote best practices in pesticide use reduction in its American potato supply.

Socially responsible investing is still in its infancy as a strategy for protecting environmental health, but it has considerable potential. By persuading people concerned about environmental health to invest in companies that are doing the right thing, SRI could make a huge difference. Like buying organic food or toxic-free consumer products, it's an example of "walking the talk" or to put it more bluntly, "putting your money where your mouth is." Indeed, if everyone concerned about environmental health invested in socially responsible companies, they'd be a tidal wave of change. Companies that protect environmental health would flourish, and those that didn't would be forced to change their ways. But support from individual investors isn't enough; to be successful SRI needs more support from the large pension funds and financial companies that control the financial markets. This is beginning to happen as more and more U.S. pension funds and mutual fund companies offer socially responsible investment options in response to client demands. This suggests that in the future SRI could pay even larger dividends as a strategy to change corporate behavior and protect environmental health.

PARTNERSHIPS WITH BUSINESS

Another strategy for changing corporate behavior is to create partnerships with progressive businesses who want to protect environmental health. Recognizing that businesses and corporations must be part of the solution, the U.S. environmental health movement is intentionally reaching out to them and engaging them in new alliances. One example is American Sustainable Business Council, a coalition of mission-driven businesses, social enterprises, and sustainable business networks working to create a just and sustainable economy. A growing network of nearly fifty business associations, representing over 150,000 businesses and 300,000 business executives, owners, investors, and others, the Council believes that U.S. society must make the transition to a sustainable and equitable economy. To do this, it has developed advocacy campaigns on corporations, election financing, energy and environment, financial markets, food and agriculture, health care, media and telecom, regulatory reform, safer chemicals, sustainable development, and taxes.

A second example is the Business-NGO Working Group for Safer Chemicals and Sustainable Materials, which was established in 2006 and is coordinated by Clean Production Action. Now comprising over eighty individuals, including business leaders from the electronics, health-care, building, apparel, outdoor industry, cleaning products, and retail sectors; investors; health-care organizations; and leaders of the U.S. environmental health movement, the Working Group supports market transitions to a healthy economy, a healthy environment, and healthy people. A third example is the Sustainable Biomaterials Collaborative, which focuses on the emerging bioplastics market. Promoting the introduction and use of plant-based products, the Collaborative seeks to replace plastics, such as PVC, that are made from fossil fuels. It also strives to create products that are sustainable from "cradle to cradle" by promoting sustainability standards, offering practical tools and case studies, and supporting policies to drive and shape the emerging markets for these products.

Developing partnerships with progressive businesses is becoming an increasingly important strategy for the U.S. environmental health movement. As the movement realizes that it can't ignore economics and that opposing capitalism and corporations isn't enough, the movement is making new friends in new places. These partnerships mark an exciting trend and could help advance environmental health in innovative and creative ways.

This chapter has examined how the U.S. environmental health movement is working to change economics, the markets, and business. It's already had considerable success with market campaigns, and its work on green chemistry and safer materials and socially responsible investing is likely to gain more traction in the future, but what stands out is the movement's strategic approach. As it becomes clearer that our economic system is destroying the planet's life support systems and threatening human health, the U.S. environmental health movement has adopted a "carrots and sticks" approach to the corporate world and is using it with increasing dexterity. On one hand, some groups attack the chemical industry head-on by calling for legislation to ban or phase out toxic chemicals. But other groups are working with the corporate world more closely than ever before. By forming partnerships and alliances with progressive companies, they're becoming cheerleaders for change. This dual approach has been used by many social movements and has always worked extremely well, so there's every reason to believe it will work to protect environmental health.

The U.S. environmental health movement's work to change economics, the markets, and business is still in its infancy, but it's very promising. By drawing attention to the costs of environmentally related diseases and disabilities, organizing market campaigns, promoting green chemistry and safer materials, encouraging socially responsible investing, and creating partnerships with progressive businesses, the movement is trying to change the entire economic system. Although there's still a long way to go, these efforts, combined with those of other social movements, are beginning to make a difference. There's also no doubt that the U.S. environmental health movement's other strategies of making environmental issues personal, advocating for precaution while making scientific arguments, and promoting environmental justice are making a difference. Not only are they protecting the health of Americans, these strategies are also changing the way people think about the environment. By pointing out that human health depends on the environment, the U.S. environmental health movement is reminding everyone that their health depends on the earth and that no one is separate from it. As Chief Seattle allegedly said, "Man did not weave the web of life; he is merely a strand in it. Whatever he does to the web, he does to himself." This realization can help to resolve the ecological crisis and prevent environmentally related illness. If we really believed that we are part of the environment, then we wouldn't do anything to harm it. In other words, we would learn to live sustainably. But for this to happen, the U.S. environmental health movement must redouble its efforts and think very strategically about how it can continue to advance social change.

NOTES

1. Sarah Anderson and John Cavanagh. Corporate empires. *Multinational Monitor* 17(12) (December 1996). Available at: http://multinationalmonitor.org/hyper/mm1296.08.html. Accessed October 31, 2012.

2. The goods and services provided "free" by nature include natural resources such as trees, fish, water, and oil and gas, as well as ecosystem functions like flood control and water purification. In 1997, leading ecological economists estimated that the value of all the goods

and services provided by nature globally was $33 trillion (Robert Costanza, Ralph d'Arge, Rudolf de Groot, et al.) The value of the world's ecosystem services and natural capital. *Nature* 387: 253–260 (1997). This is more than the value of traded goods and services in that year.

3. As quoted in: E. Johnson. *Charles Dickens: His tragedy and triumph.* Vol. 2, p. 715. New York, NY: Simon Schuster (1952).

4. World Health Organization. *Preventing disease through healthy environments: Towards an estimate of the environmental burden of disease.* Geneva, Switzerland (2006). Available at: http://www.who.int/quantifying_ehimpacts/publications/preventingdisease. Accessed October 31, 2012.

5. Joel Schwartz, Hugh Pitcher, Ronnie Levin, Bart Ostro, and Albert L. Nichols. *Costs and benefits of reducing lead in gasoline: Final regulatory impact analysis.* EPA-230/05-85/006. Washington, DC: U.S. Environmental Protection Agency (1985).

6. David S. Salkever. Updated estimates of earnings benefits from reduced exposure of children to environmental lead. *Environmental Research* 70(1): 1–6 (1995).

7. Scott D. Grosse, Thomas D. Matte, Joel Schwartz, and Richard J. Jackson. Economic gains resulting from the reduction in children's exposure to lead in the United States. *Environmental Health Perspectives* 110(6): 563–569 (2002).

8. Elise Gould. Childhood lead poisoning: Conservative estimates of the social and economic benefits of lead hazard control. *Environmental Health Perspectives* 117(7): 1162–1167 (2009).

9. Leonardo Trasande, Philip J. Landrigan, and Clyde Schecter. Public health and economic consequences of methyl mercury toxicity to the developing brain. *Environmental Health Perspectives* 13(5): 590–596 (2005).

10. Lars Hylander and Michael Goodsite. Environmental costs of mercury pollution. *Science of the Total Environment* 368: 352–370 (2006).

11. Ben H. Carpenter, James R. Chromy, Walter D. Bach, D. A. LeSourd, and Donald G. Gillette. Health costs of air pollution: A study of hospitalization costs. *American Journal of Public Health* 69(12): 1232–1241 (1979).

12. Nicolas Mueller, Robert Mendelsohn, and William Nordhaus. Environmental accounting for pollution in the United States economy. *American Economic Review* 101: 1649–1675 (2011).

13. Dirty air costs California economy $28 billion annually. California State University, Fullerton. News and Information, November 12, 2008. No. 091. Available at: http://calstate.fullerton.edu/news/2008/091-air-pollution-study.html. Accessed October 31, 2012.

14. Perry Sheffield, Angkana Roy, Kendrew Wong, and Leonardo Trasande. Fine particulate matter pollution linked to respiratory illness in infants and increased hospital costs. *Health Affairs* 30(5): 871–878 (2011).

15. Patrick Kinney and Amruta Nori-Sarma. *Health and economic benefits of clean air regulations: White Paper.* Joint Center for Political and Economic Studies, Washington, DC. (2011). Available at: http://www.jointcenter.org/research/white-paper-health-and-economic-benefits-of-clean-air-regulations. Accessed October 31, 2012.

16. Kim Knowlton, Miriam Rotkin-Ellman, Linda Geballe, Wendy Max, and Gina Solomon. Six climate change-related events in the United States accounted for about $14 billion in lost lives and health costs. *Health Affairs* 30(11): 2167–2176 (November 2011).

17. Philip J. Landrigan, Clyde B. Schechter, Jeffry M. Lipton , Marianne C. Fahs, and Joel Schwartz. Environmental pollutants and disease in American children: Estimates of morbidity, mortality, and costs for lead poisoning, asthma, cancer, and developmental disabilities. *Environmental Health Perspectives* 110(7): 721–728 (2002).

18. Rachel Massey and Frank Ackerman. *Costs of preventable childhood illness: The price we pay for pollution.* Medford, MA: Global Development and Environment Institute, Tufts University (2003).

19. Kate Davies. Economic costs of childhood diseases and disabilities attributable to environmental contaminants in Washington state. *Eco Health* 3: 86–94 (2006).

20. Oregon Environmental Council. *The price of pollution: Cost estimates of environmentally-related disease in Oregon.* (2008). Available at: http://www.oeconline.org/resources/publications/reportsandstudies/pop. Accessed October 31, 2012.

21. Leonardo Trasande and Yinghua Liu. Reducing the staggering costs of environmental disease in children estimated at $76.6 billion in 2008. *Health Affairs* 30(5): 863–870 (May 2011).

22. *EPA fact sheet: Mercury and air toxics standards.* (2011). Available at: http://www.epa.gov/mats/pdfs/20111221MATSimpactsfs.pdf. Accessed October 31, 2012.

23. Laura Terragni. From the White Label campaign to the no sweat initiatives, a journey at the roots of political consumerism. Proceedings of the Nordic Consumer Policy Research Conference 2007. Available at: http://www.docstoc.com/docs/39156652/PROCEEDINGS-of-the-Nordic-Consumer-Policy-Research-Conference. Accessed October 31, 2012.

24. Baby Milk Action. *Briefing paper: History of the campaign.* Undated. Available at: http://www.babymilkaction.org/pages/history.html. Accessed October 31, 2012.

25. Ideas for action: The baby milk issue. *New Internationalist Magazine* Issue 110 (April 1982).

26. World Health Organization. *International code of marketing breast-milk substitutes.* (1981). Available at: http://www.who.int/nutrition/publications/code_english.pdf. Accessed October 31, 2012.

27. Consumer Electronics Association. *U.S. consumer electronics sales & forecasts* . Available at: http://www.ce.org/Research/Products-Services/Industry-Sales-Data.aspx. Accessed October 31, 2012.

28. Apple's worldwide headquarters are located in Cupertino, California.

29. Greenpeace International. *Green my Apple bears fruit.* Available at: http://www.greenpeace.org/international/news/greening-of-apple-310507. Accessed October 31, 2012.

30. Greenpeace International. *Green my Apple bears fruit.* Available at: http://www.greenpeace.org/international/news/greening-of-apple-310507. Accessed October 31, 2012.

31. Steve Jobs. *A greener Apple.* May 2, 2007. Available at: http://www.apple.com/hotnews/agreenerapple/. Accessed October 31, 2012.

32. United Nations Environment Programme. *Basel Conference addresses electronic waste challenge.* November 27, 2006. Available at: http://www.unep.org/documents.multilingual/default.asp?DocumentID=485&ArticleID=5431&l=en. Accessed October 31, 2012.

33. U.S. EPA. *Statistics on the management of used and end-of-life electronics.* Available at: http://www.epa.gov/osw/conserve/materials/ecycling/manage.htm. Accessed October 31, 2012.

34. United Nations Environment Programme. *Urgent need to prepare developing countries for surge in e-wastes.* February 22, 2010. Available at: http://www.unep.org/Documents.Multilingual/Default.asp?DocumentID=612&ArticleID=6471. Accessed October 31, 2012.

35. HR 2284. To prohibit the export from the United States of certain electronic waste, and for other purposes. 112th Congress 1st Session. Available at: http://www.gpo.gov/fdsys/pkg/BILLS-112hr2284ih/pdf/BILLS-112hr2284ih.pdf. Accessed October 31, 2012.

36. S 1270. To prohibit the export from the United States of certain electronic waste, and for other purposes. 112th Congress 1st Session. Available at: http://www.gpo.gov/fdsys/pkg/BILLS-112s1270is/pdf/BILLS-112s1270is.pdf. Accessed October 31, 2012.

37. Whitfield & Associates. *The economic benefits of polyvinyl chloride in the United States and Canada.* Prepared for the Chlorine Chemistry Division of the American Chemistry Council and the Vinyl Institute. (December 2008). Available at: http://www.pvc.org/upload/documents/The_Economics_of_PVC.pdf. Accessed October 31, 2012.

38. *Sam Suds and the case of PVC, the poison plastic.* (2006). Video available at: http://chej.org/campaigns/pvc/projects/sam-suds/. Accessed October 31, 2012.

39. Target will reduce PVC use. *New York Times.* November 6, 2007. p. D2.

40. Centers for Medicare and Medicaid Services, Office of the Actuary. *National health statistics group.* Available at: http://www.cms.hhs.gov/NationalHealthExpendData. Accessed October 31, 2012.

41. Paul Anastas and John Warner. *Green chemistry: Theory and practice* . Oxford University Press: New York (1998).

42. Pike Research. *Green chemistry.* (2011). Available at: http://www.pikeresearch.com/research/green-chemistry. Accessed October 31, 2012.

43. Greenwashing is the act of misleading consumers about the environmental practices of a company or the environmental benefits of a product or service.

44. See Clean Production Action at: http://www.cleanproduction.org. Accessed October 31, 2012.

45. The Forum for Sustainable and Responsible Investment. *Report on sustainable and responsible investing trends in the United States, 2012*. Available at: http://ussif.org/resources/pubs/. Accessed November 14, 2012.

46. Social Investment Forum. *Socially responsible investing facts*. Available at: http://www.socialinvest.org/resources/sriguide/srifacts.cfm. Accessed October 31, 2012.

47. Gary B. Nash and Jean R. Soderlund. *Freedom by degrees: Emancipation in Pennsylvania and its aftermath*. New York: Oxford University Press (1991).

48. John Wesley. *The use of money, sermon 50* (1872).

49. *CERES principles*. Available at: http://www.ceres.org. Accessed October 31, 2012.

Conclusion and Next Steps: Strategies for Social Change

In their final major assignment, students at Antioch University Seattle's Center for Creative Change—where I teach—are asked to present their ideas about social change. Whether they're getting a master's degree in Environment & Community, Organizational Development, or Whole Systems Design, all graduating students are required to discuss the strategies they can use to advance social change wherever they work, or would like to work. Specifically, students are asked to reflect on the experiences of social change they acquired during a nine-month hands-on "change project" and to synthesize this applied learning with current theories and concepts about change. This experiential approach to learning is very powerful. Based on the belief that social change cannot be taught, it requires learning that is grounded in the real world. By learning from their own experiences—their failures as well as their successes—students develop the practical knowledge and skills they need to advance social change.

Learning from experience is critically important for social movements too. Unless they can reflect on what they did, and whether or not their actions were effective, they probably won't be able to design successful strategies for the future. So what can the U.S. environmental health movement learn from its experiences of social change over the past thirty-five years? What can it learn about social change from other movements? And how can it use this learning to design strategies for the future? This concluding chapter describes some possible responses to these questions.

STRATEGIES FOR SOCIAL CHANGE

Over its lifetime, the U.S. environmental health movement has employed several overarching and well-established strategies for social change. Described in the preceding chapters of this book, they include organizing collective action on locally identified issues, lobbying for new legislation, and building national coalitions. Although the use of these strategies is more complex and nuanced than can be described in this book, there are several discernible trends.

Organizing collective action on locally identified issues was the principal strategy used by the first environmental health groups formed in the late 1970s and early 1980s, such as the Love Canal Homeowners' Association. These local groups, many of whom were oriented primarily towards environmental justice, relied on grassroots organizing and direct action to protest toxic waste dumps and other sources of pollution in their neighborhoods. Community members identified local problems, talked with each other, held public meetings, signed petitions, organized demonstrations, and got media attention to clean up pollution and prevent the construction of new hazardous facilities. This collective action approach has been used by most social movements; indeed, it's fundamental for social change.

The movement's second strategy—lobbying for new legislation—became more prominent in the late 1980s as state groups started to advocate for stronger legislation. Identifying policy issues based on what they thought was winnable, they've been very effective in passing state and local legislation on toxic chemicals. However, the increasing use of this approach created tension with some local environmental health and justice groups. The sources of this tension were threefold. Local groups, many of which were located in poor communities or communities of color, felt increasingly ignored and marginalized by state groups because of their focus on winnable legislation, rather than locally identified problems. To add to this, local groups were mostly run by passionate but penniless volunteers who felt that professionally managed state and national groups were scooping up most of the available foundation funding. And finally, the lobbying approach used by state groups represented a very different strategy for achieving social change. Unlike local collective action and protest, it was based on political and legal mediation and appeared soft and accommodating to angry grassroots protestors. For these reasons, a certain wariness emerged between local environmental health groups on one hand and state ones on the other.

The need for the third strategy—building national coalitions—became obvious in the early 2000s, when state groups realized that they could be even more effective by coordinating their efforts across the country and reaching out to new constituencies, including people affected by environmentally related diseases, the organizations that represent them, and local groups. This realization led to the formation of diverse national coalitions such as *Safer Chemicals, Healthy Families*, as well as to national and state groups playing a larger role in coordinating and supporting local groups. This strategy is essential for the movement's future growth. By building diverse coalitions and strengthening its internal connections, the U.S. environmental health movement could become even more successful. For instance, greater efforts to develop alliances with organizations working on public health and occupational health could be helpful. Their emphasis on advancing social justice by preventing disease makes them natural allies for the environmental health movement. Coalitions are harder to disrupt and defeat because the opposition must deal with many different types of people and groups, as well as a decentralized power structure. Like a spider's web, the future strength of the U.S. environmental health movement lies in its connecting threads and overall design.

Today, these three strategies are converging, and there's a growing recognition that they're all necessary for social change. For instance, as well as organizing collective action, local groups are passing ordinances, home rule charters, and other local legal instruments to control polluting industries within their boundaries (see chapter 5).

But in addition to these overarching and well-established strategies, the U.S. environmental health movement uses other unique strategies to advance social change, including making environmental issues personal (see chapter 6), making its case using science and precaution (see chapter 7), arguing for environmental justice (see chapter 8), and trying to change the economic system (see chapter 9). Combined with organizing collective action at the local level, lobbying for new legislation, and building national coalitions, these unique strategies are helping to safeguard environmental health.

Superficially, it might appear that the most successful of these approaches is passing state and local legislation. After all, between 1990 and 2009 over nine hundred state and local toxics policies were proposed or enacted, and between 2003 and 2011 eighteen states passed seventy-one chemical safety laws (see chapter 5), but I would argue that the movement's efforts to put a human face on environmental issues and confront environmental injustice are more significant. These two strategies are not only protecting environmental health, they've transformed public discourse on the environment. Because they reach beyond intellectual and scientific ways of knowing and speak in a deeply personal, ethical, and spiritual way, they touch people's hearts and souls.

Taken together, all of these strategies are changing the way Americans think about the environment. The U.S. environmental health movement is making people realize that the environment is "in here." It's not just something "out there" that can be controlled and manipulated; it's also part of us and we are part of it. As we come to understand that the same toxic chemicals that are in air and water are also present in our blood and bones and that our health is affected just like wildlife health, it becomes impossible to maintain the illusion that human beings are separate from and superior to nature. By changing this anthropocentric belief, the U.S. environmental health movement is changing the fundamental values of Western culture, and in doing so it offers a way to resolve the ecological crisis. Although the U.S. environmental health movement still has a long way to go, it is making a difference.

But to achieve environmental health for all, the movement needs to become even more strategic in its thinking. Its ability to design effective strategies has already helped, but as the ecological and environmental health crises worsen, it must think even more carefully about how to navigate a path into the future. To do this, it could learn a lot from previous social movements.

So what can the U.S. environmental health movement learn from previous social movements? I believe that there are many lessons, including the importance of creating inspiring visions; "minding the gap" between our collective aspirations and reality; seeing the forest and the trees; identifying leverage points for environmental health; organizing more collective action; telling environmental health stories; and self-care. It's not that the movement is not using these strategies, but it could use them more intentionally.

CREATING INSPIRING VISIONS

The history of social movements shows that inspiring, idealistic visions provide a powerful impetus for social change. Just think of the civil rights movement, the peace movement, and the women's movement. They all used core American ideals to construct simple, but compelling narratives about the type of society they wanted to create. In fact, no social movement has ever succeeded by being only negative and critical; it's just too depressing. People need to see themselves not only as victims of unjust systems and institutions, but also as having the power to change them.

Inspiring visions are important because they provide a sense of empowerment and agency and affirm that another world is possible. Offering a direction and a purpose for social change, idealistic visions draw people forward and encourage them to take action to realize their common dreams and aspirations. In doing so, they can attract new participants to join movements and help to keep them engaged. By supporting movement members along the often slow and arduous path of social change, inspiring visions provide the psychological and spiritual sustenance to keep going even when things look bleak. They can also unify a social movement by spelling out its shared values and beliefs. Without an unambiguous declaration of what a movement stands for, it's easy to forget why it exists and fritter away scarce resources on insignificant issues. Providing the spark that can transform despair into hope, inspiring idealistic visions can help to ignite collective action and promote large-scale social change.

One of the most famous is, of course, Martin Luther King Jr.'s 1963 "I have a dream" speech. Symbolically delivered on the steps of the Lincoln Memorial, one hundred years after President Abraham Lincoln issued the Emancipation Proclamation, this speech is justifiably celebrated for laying out a bold vision for the civil rights movement and the entire United States. What made it so remarkable is that it drew on the secular ideals of freedom, equality, and justice expressed in the Declaration of Independence, the Constitution, and the Emancipation Proclamation and then put them in a spiritual context by arguing that racial justice is in accord with God's will. In doing so, this speech offered redemption to white Americans for the sin of racial discrimination.[1]

The importance of inspiring visions for social change should not be a surprise. Visions embodying ideals such as freedom, justice, love, equality, and peace have always helped humankind cope with life's hardships and disappointments. They speak to our highest hopes for the future and resonate deeply within us. Indeed, without the possibilities engendered by positive visions for the future, humankind would probably not survive. As the Bible says: "Where there is no vision, the people perish."[2] My favorite quotation on this topic is more secular and comes from *Alice in Wonderland*:

> "Would you tell me, please, which way I ought to go from here?" said Alice.
> "That depends a good deal on where you want to get to," said the Cat.
> "I don't much care where," said Alice.
> "Then it doesn't matter which way you go," said the Cat.[3]

Like Alice, some environmental health activists don't seem to know where they want to go. For them, it's enough to denounce corporate polluters and criticize ineffective government legislation. They don't seem to have a clear positive vision of what they're working for; it's enough to be against the *status quo*. This is very understandable. Corporations have been getting away with polluting the environment for decades, and government agencies haven't done much to stop them. Given this situation, it's only natural to criticize and oppose. One activist told me, "When people think their health and the health of their children is being attacked, of course they're going to fight back."[4] Defending oneself and one's family against environmental health threats is an instinctive response.

But the U.S. environmental health movement should do more than just criticize and oppose; it should also create inspiring idealistic visions that describe the type of society it wants to create. What would environmental health actually look like? What are the values, beliefs, and characteristics that underlie environmental health? How would we know when we've achieved environmental health? To be empowering and engaging, the movement must address these and similar questions. This isn't to suggest that it should gloss over the harsh realities of environmentally related diseases and disabilities or paint a naively optimistic Pollyanna-ish picture of the future. But the movement should do more to speak to Americans' collective ideals and aspirations. It ignores the need for inspiring idealistic visions at its peril.

At this point, I'd like to acknowledge that the movement is already offering many positive alternatives. Its work on green chemistry and safer materials and community-based economic development and its partnerships with progressive businesses demonstrate that it's extremely serious about developing constructive solutions. But this work is based on individual issues and doesn't present visions for environmental health as a whole.

Given the magnitude of the ecological and environmental health crises, articulating inspiring visions for the future may appear to be a challenging task, but it's not impossible. After all, many previous movements related to environmental health have eloquently expressed their hopes and dreams. In the nineteenth century, social reformers and sanitarians articulated a positive vision for public health which paved the way for the government measures that led to the prevention of environmentally transmitted diseases. And in the early twentieth century, the Progressive Era's idealism about conserving the environment for future generations led to the first environmental protection legislation, and the creative ideas of the City Beautiful and Garden City movements for improving the quality of life in towns and cities resulted in new urban forms (see chapter 2). By drawing on these examples, the U.S. environmental health movement should be able to create its own inspiring visions for the future.

Health itself is a compelling collective value that the movement hasn't sufficiently exploited. Drawing attention to health as a social ideal could be very useful because everyone wants to be healthy. Cutting across social, political, economic, and cultural divisions, it's one of the values shared by all human beings. Health is a powerful ideal for several different reasons. First, people value their health because it contributes to their overall well-being and happiness. In this way, it's an intrinsic good—something enjoyed and desired for its own sake. But even if people aren't consciously aware of being healthy, when they lose their health they recognize its importance. In this way, health is an instrumental good that enables us to live our lives and participate in society; in other words, it's a resource for living.[5] Health is also a social good because it contributes to national productivity through increased participation in the workforce and reduced absenteeism. By taking advantage of the fact that health is a core American value, the U.S. environmental health movement could advance its cause.

MINDING THE GAP BETWEEN OUR COLLECTIVE ASPIRATIONS AND REALITY

But even though the U.S. environmental health movement should develop inspiring, idealistic visions, it shouldn't abandon its critical, oppositional stance. Not at all. In fact, another lesson from previous social movements is that drawing attention to the gap between the type of society people want and the type of society they actually have is a very powerful strategy for advancing social change. This was another brilliant feature of King's "I have a dream" speech. It contained both a positive picture of the future and an incisive critique of the racism and discrimination that pervaded the United States in the 1960s. I call this two-pronged approach "minding the gap."

Many train and subway stations warn the public to mind the gap between the train doors and the platform. And if they don't, would-be passengers can end up falling flat on their faces. Similarly, movements that fail to mind the gap between aspirational social values and the reality of everyday life risk falling on their faces. It's not enough to offer positive, inspiring visions of the future or to present a negative, critical analysis of what's wrong; social movements need to do both at the same time, thereby drawing attention to the gap between them.

Public opinion surveys show that most people want the same things—health and well-being; a clean, safe environment; a job that pays a living wage; and strong, vibrant communities. But the truth is that many suffer from illness; our air, water, and soil are contaminated; most people's disposable income is stagnant or declining; and many communities are struggling for survival. Quite simply, there's a yawning chasm between our collective aspirations and reality, and successful social movements highlight this fact. In the 1960s, the civil rights movement contrasted the belief that "all men are created equal" with the segregation, discrimination, and injustice experienced by African Americans. In the 1970s, the environmental movement highlighted the disparity between the ideal of "America the Beautiful" with the reality of pollution and waste. And in the 1980s, the antinuclear movement emphasized the hypocrisy of trying to safeguard world peace by maintaining huge stockpiles of nuclear weapons.

This strategy works because it introduces a cognitive dissonance that makes people very uncomfortable. The awareness that society operates in ways that conflict with our most deeply held aspirations and ideals can be extremely distressing. We like to think that our social institutions and systems are consistent with our shared values, so it's very difficult to live with the realization that they aren't. And when this collective discomfort becomes too great, we try to resolve it so we can feel better about ourselves. One of the ways of doing this is by changing our societal institutions and systems, so they're more consistent with what we want. This is the essence of progressive social change. By combining positive, idealistic visions with incisive social criticism and drawing attention to the gap between them, social movements can stimulate social change.

For the U.S. environmental health movement, this could mean highlighting the gap between core American values, such as health, happiness, equality, and family, with the reality of environmentally related diseases and disabilities. Complementing the work of environmental justice activists who already highlight the gap between the core American value of justice and the disproportionate risks faced by marginalized communities and populations, this strategy could help to advance environmental health.

SEEING THE FOREST AND THE TREES

Successful movements also have a capacity to see the whole forest and the individual trees at the same time—keeping a wide-angle view on social change even as they focus on specific issues. For instance, achieving gender equality was the central issue for the women's movement, and it advanced this cause by working on several smaller issues including reproductive and abortion rights, equal pay, and career opportunities for women. Leaders of the women's movement understood that work on individual issues was necessary to advance its central concern, and that its central concern supported work on individual issues. They kept their eyes on the forest as well as the trees. But it's often easy for activists to get completely absorbed in detailed campaigns and to lose sight of the big picture. This isn't helpful because it can dissipate a movement's energy and obscure its overall goals. Moreover, getting lost in individual issues serves the interests of the power-holders. If they can keep activists focused on the details, they can prevent social movements from coalescing and gaining strength. The power-holders understand that strength comes from numbers, so by keeping activists focused on individual issues they can prevent social movements from becoming powerful. For these reasons, being able to see the forest and the trees at the same time is an important skill.

So far, the U.S. environmental health movement has focused on toxic chemicals, and although this is a critically important issue, it's not the only one. To become a true environmental health movement, it will need to embrace a broader range of issues. Work on nanotechnology, EMFs, fossil fuels, food, green building, and other issues is a good start, but there are many, many other factors that influence environmental health. If the movement wants to be seen as more than just an antitoxics movement, it should consider the whole forest of environmental health, not only the tree of toxic chemicals.

This is beginning to happen as the U.S. environmental health movement explores the concept of ecological health. Extending earlier work on the determinants of health[6,7,8] and ecosystem health,[9,10] ecological health recognizes the interrelationships among all the factors that affect health—social, economic, personal, genetic, and environmental—as well as the dependence of human health on the planet. Ted Schettler and the Science and Environmental Health Network were among the first to apply the concept of ecological health to environmental health, although they used the term "ecological medicine."[11]

Whether it's called ecological health or ecological medicine, this concept takes a systems view of environmental health and addresses the problem of proving causality (see chapter 7). By pointing out that environmental health effects are caused by the interaction of many social, economic, and environmental factors—what could be called a web of causality—ecological health transforms so-called confounding factors, such as race, ethnicity, and economic status, into "effect modifiers."[12] Rather than being factors to be "controlled" or eliminated from scientific studies, effect modifiers complexify and add to understanding. For instance, there's growing evidence that the health effects of toxic substances can be modified by nutritional status,[13] and many agree that the epidemic of obesity is likely caused by multiple interacting influences, including environmental, metabolic, genetic, behavioral, dietary, social, and psychological factors. By drawing attention to these interactions and connections, ecological health transcends simplistic notions of cause and effect in environmental health science.

In 2007, the concept of ecological health led to the development of *A Common Agenda for Health and Environment.*[14] Under the auspices of the University of Massachusetts at Lowell and the Lowell Center for Sustainable Production, nearly one hundred environmental health leaders gathered to craft an extremely ambitious and wide-ranging plan to achieve environmental health within a single generation. Going well beyond toxic chemicals, it advocates for many issues not previously regarded as environmental health concerns, including promoting energy security, developing a national blueprint for biodiversity preservation, and establishing a Department of Peace.

These ideas about ecological health indicate that the U.S. environmental health movement is beginning to see the entire forest, as well as individual trees. Having this type of overview will allow the movement to identify strategic leverage points to advance its cause.

IDENTIFYING LEVERAGE POINTS FOR ENVIRONMENTAL HEALTH

Leverage points are places in complex systems where a small shift in one thing can produce big changes across an entire system. In other words, they're points of power.[15] Identifying leverage points requires understanding the whole system you're trying to change, especially the interactions between the component parts, the trends and patterns across the system, and the most important influences on system behavior. Then, by identifying and working on issues that will change the entire system's behavior, one can leverage major change. This strategy has been used by many social movements. For instance, early civil rights leaders identified voting rights as a key leverage issue because it spoke to the public—white and African American—in terms of the American ideals of democracy and citizenship, and more importantly, it had the potential to change the composition of U.S. political institutions and governance structures. This combination made it an excellent leverage issue for social change.

As the U.S. environmental health movement moves beyond being just an antitoxics movement, it could identify leverage points for the entire system of environmental health. Although the movement has identified and worked on leverage points for toxic chemicals, such as market campaigns, it hasn't yet identified many points that affect the entire environmental health system. One of them may be health-care costs. The United States has the most expensive health-care system in the world, and it's getting more expensive all the time. In 2013, U.S. health expenditures are expected to be in excess of $2.9 trillion, and by 2021 they are expected to rise to almost $4.8 trillion, about one-fifth of the U.S. gross domestic product.[16] But it's also one of the least effective health-care systems among industrialized countries. The *World Health Report 2000*[17] ranked the U.S. health-care system thirty-seventh in the world, and in terms of health outcomes, in 2006 the United States ranked thirty-ninth for infant mortality, forty-third for adult female mortality, forty-second for adult male mortality, and thirty-sixth for life expectancy.[18]

There are many reasons for this poor performance. One of them may be that the U.S. system is treatment oriented and does very little to prevent illness, despite the fact that many expensive and common chronic diseases are preventable. Indeed, it's been estimated that an investment of ten dollars per person per year in disease prevention programs could significantly improve public health and yield net savings of nearly $18 billion annually in ten to twenty years (in 2004 dollars).[19] And these estimates don't include the significant economic gains that could be achieved as a result of increased worker productivity, reduced absenteeism at work and school, and a better quality of life.

For these reasons, health care costs could be an important leverage issue for the U.S. environmental health movement. Studies on the costs of environmentally related diseases and disabilities (see chapter 9) are a good start, but additional work could further exploit this issue. By making a case that investments in environmental quality could significantly reduce health costs, improve worker productivity, and enhance population health, the movement could advance its cause. Linking environmental health to dollars and cents is likely to be an effective strategy for social change because economics is so important in today's world. In fact, the economics of environmental health may be the single most important leverage point in the struggle for environmental health.

Another leverage point may be climate disruption—"one of the most serious public health threats facing our nation" according to the president of the American Public Health Association.[20] Similarly, a recent draft report from the U.S. Global Change Research Program stated "Climate change will influence human health in many ways; some existing health threats will intensify, and new health threats will emerge. . . . In fact, U.S. population growth has been the greatest in coastal zones and in the arid Southwest, areas that already have been affected by increased risks from climate change."[21] With the potential to cause unprecedented effects on human health and well-being, climate disruption could be an extremely powerful place for the environmental health movement to intervene. But it has not really been framed as a serious health issue yet. By focusing on quantitative scientific data about the environmental effects of climate change, environmentalists have failed to make this obvious connection. For instance, activist Bill McKibben's landmark 2012 article on *Global Warming's Terrifying New Math*[22] and his *Do the Math*[23] campaign are full of facts and figures about climate disruption, but they don't mention health. This is very strange. Despite the abundant scientific evidence of health effects and the fact that most Americans now believe in climate disruption because of their personal experiences of extreme weather events,[24,25] the health argument has not been used to any great extent and climate disruption still appears abstract and remote to many. This creates an excellent opportunity for the U.S. environmental health movement.[26] By personalizing the effects of climate disruption and drawing attention to its health effects, it could leverage social change—not only on this issue but on other related issues, such as toxic chemicals manufactured from petroleum feedstocks, water quality and use, and agriculture.

ORGANIZING MORE COLLECTIVE ACTION

Although I've already discussed collective action on locally identified issues as one of the U.S. environmental health movement's strategies, I want to talk about it some more because it's so important for social change. Just think about the nonviolent "people power" revolution in the Philippines in 1986 that overthrew the repressive regime of President Ferdinand Marcos, or the mostly peaceful popular uprisings in Eastern Europe in the late 1980s that led to the fall of the Berlin Wall and the dissolution of the Soviet Union. Successful social movements have always had strong support from ordinary people who have been willing to engage in civil disobedience to protest *en masse*.

Collective, peaceful civil disobedience has been an essential strategy for U.S. social movements too. Would Congress have passed the *Civil Rights Act* (1964) or the *Voting Rights Act* (1965) without the civil rights movement's mass marches, sit-ins, and boycotts? Would the labor movement have been successful in securing fair wages and benefits for its members without large-scale demonstrations? Would President Richard Nixon have ordered the withdrawal of troops from Vietnam without the hundreds of protests organized by the peace movement across the country? Probably not. Collective, peaceful civil disobedience may be a necessary strategy for social change.

The U.S. environmental health movement already enjoys support from millions of ordinary Americans. Indeed, this is one of its greatest assets. As one activist told me "Our strength isn't measured in dollars—we don't have high priced lobbyists or political expense accounts. But we've got the public on our side, and that is the source of our strength. No amount of money can buy that kind of political clout."[27] But although its local groups use collective action, as a whole, the movement rarely translates its public support into visible demonstrations of protest. Despite the fact that it has sophisticated electronic communication networks, they are infrequently used to mobilize the public onto the streets. Perhaps this is because they think that people are more fearful of participating in mass demonstrations than they used to be or because state and national groups are more focused on political negotiation and compromise, but whatever the reason many national and state groups seem to downplay the importance of collective, peaceful civil disobedience as a strategy for social change.

It is true that the U.S. environmental health movement has mobilized its supporters in market campaigns (see chapter 9), some of which make the issues amusing and fun. Although this can be effective, it's important to remember that environmental health problems are deadly serious and are unlikely to be resolved by throwing a light-hearted party. Other market campaigns ask consumers to change their purchasing habits or use electronic media to urge corporations to change their ways. But these approaches encourage individual action; they aren't really about taking visible collective action *en masse*. They don't require movement supporters to take personal risks to show their commitment to the cause, and they don't provide physical evidence of the movement's strength.

To become a serious resistance movement, the U.S. environmental health movement should consider collective, peaceful civil disobedience more often. Like the suffragettes who used it in their struggle for the right to vote or the African Americans who used it to protest for their rights to equal treatment, movement supporters may need to publicly show the strength of their convictions more often.

TELLING ENVIRONMENTAL HEALTH STORIES

Another strategy used by social movements is encouraging ordinary people to tell their personal stories. Storytelling has always brought people together and helped them to make sense of life's ups and downs. Because it's a lens through which to process experiences and make meaning of life, storytelling can be a very effective strategy for social change. By telling their stories and listening to those of others, people come to appreciate that they share similar experiences and that no one is alone, and as a result, they develop shared ideas that can lead to collective action and demands for change.

This strategy has been used by many previous social movements. For instance, in the 1970s and 1980s, the women's movement relied on storytelling to build its grassroots base and develop its theory of change. Coming together in kitchens and living rooms across the country, middle-class wives and mothers shared their stories in what were called "conscious raising groups." The similarities among the women's experiences quickly became obvious, and they used them to construct a theory about gender inequality and discrimination. By listening to each other, they saw links between their experiences, including the undervaluing of work in the home, the lack of employment rights (especially for married women), the lack of reproductive rights, and unequal access to educational opportunities for girls and young women. Growing directly out of their lived experiences, their theory of social change was grounded in reality, rather than abstract academic concepts.

Two of the leading proponents of storytelling as a strategy for social change in the twentieth century were Myles Horton, founder of the Highlander Folk School (now the Highlander Research and Education Center) in Tennessee, and Brazilian educator Paulo Freire. Describing this approach, Horton said, "The one thing they know is their own experience. . . . They want to talk about their own experience. Then other people join in and say, 'Ah ha, I had an experience that relates to that.' So pretty soon you get everyone's experiences coming in, centered around that one person's experience."[28] Brazilian educator Paulo Freire used a similar strategy. By encouraging peasants living in Brazil's slums to talk about their experiences of oppression, he helped them to break through what he called "the culture of silence." This made them realize that they all shared similar experiences and that by working together they could bring about social change.[29]

The U.S. environmental health movement is learning to use this strategy. Its approach of making environmental issues personal (see chapter 6) highlights the stories of people affected by environmentally related disease, and several groups' web sites feature personal stories. In addition, the movement has a few extremely eloquent storytellers. Key among them is Sandra Steingraber, whose three best-selling books[30, 31, 32] and numerous articles on environmental health exemplify the art. One reviewer declared, "What she's brilliant at—almost in a league of her own—is mixing personal passionate stories with totally comprehensive and accurate science."[33] Steingraber touches readers' hearts as well as their heads, and in doing so she engages people in a way that scientific data alone cannot. The abstract becomes real, the objective becomes subjective, and the personal becomes political. Blending science with lyrical memoir, Sandra Steingraber's books and articles have captivated, educated, and inspired countless readers to take action on environmental health.

But although the stories of a few individuals can be very inspiring, this strategy is about much more. It's about encouraging ordinary people to come together to tell their stories and then use them to inspire grassroots activism. To use this strategy more effectively, the U.S. environmental health movement could broaden its understanding of how social change happens. Many groups, especially at the state and national levels, believe that change will happen when people know the scientific facts about how environmental quality affects health, so their social change strategy is based on providing information and education. Underlying this approach is a belief that people are empty vessels who need to be motivated by filling them up with facts and figures. Storytelling flips this upside down and begins with what people already know. By encouraging ordinary people to talk about their stories and find meaning in their lives, storytelling supports and validates what they already know; it starts where they are. This could be a powerful strategy for the U.S. environmental health movement. By building on people's experiences of environmental health problems, it could support and empowers them to take collective action. Of course, storytelling will never replace scientific information—nor should it. Environmental health should always take account of the science, but motivating collective action for environmental health is also likely to require the more personal approach of telling stories.

SELF-CARE

A final lesson from previous social movements is self-care. As author and activist Parker Palmer says: "self-care is never a selfish act—it is simply good stewardship of the only gift I have, the gift I was put on earth to offer to others. Anytime we can listen to true self and give it the care it requires, we do so not only for ourselves but for the many others whose lives we touch."[34] Working on environmental health is often exhausting, painful, and discouraging, and all too many activists get burned out, depressed, or physically ill. Caught up in the relentless social pressure to keep busy, they don't look after themselves. This is bad for the individuals concerned and bad for the environmental health movement. Although work on social change is important and urgent, environmental health activists must take time to look after themselves. Indeed, how can we say we want to protect the public's health, if we fail to look after our own?

Recognizing that real social change is never easy or quick, environmental health activists should prepare for the long haul. Because the work is often challenging, they need to pay attention to their own health. So how can environmental health activists look after themselves? I believe that there are several ways, including reaching outwards to build supportive communities, turning inwards to develop our individual power, and reconnecting with the earth. Using these practices, activists can build the personal resources needed to sustain their efforts over the long term.

Reaching out to families, friends, colleagues, and other like-minded people and building supportive and caring communities can be very nourishing. Indeed, having strong social networks is an important determinant of health.[35] When we're feeling discouraged, exhausted, or burned out, relationships with others can be sustaining. Conversely, when we're feeling strong and confident, we can help others who are feeling low. In short, we need other people and they need us. The English metaphysical poet John Donne (1572–1631) was well aware of this when he wrote, "No man is an island, entire of itself."[36]

A second strategy for self-care is to turn inwards. This doesn't mean becoming a hermit or a monk, but it does mean trying to make time to cultivate our internal strength. This can involve meditating, praying, or just pausing to develop a sense of stillness amid the busyness of daily life. By developing our personal power from the deepest places within ourselves, turning inwards helps to develop the capacities that activists need to do their work in the external world.

Reconnecting with the earth—the very ground of our being—is a third strategy for self-care, especially for people working on environmental health. We all come from the earth and will return to it when we die. It feeds our bodies and nourishes our hearts and souls. But how many activists make time to draw strength from the living earth on a regular basis? Reconnecting with nature can be a powerful way of revitalizing ourselves. In my own life, I know that being in nature helps me to be more fully alive and present in the world.

As the ecological and environmental health crises continue to worsen and the need for social change becomes more urgent, it becomes even more important for movement activists to look after themselves. In doing so, we will be able to see things as they really are—however difficult, painful, and overwhelming—act more decisively, and become more effective agents for social change.

FINAL REFLECTIONS

As the U.S. environmental health movement navigates a path into the future, it must always remember its central goal—environmental health for all. This goal will require changing our culture's social, economic, and political systems and institutions, as well as the values and beliefs on which they are based. Social change of this magnitude is not about passing one piece of legislation or another, cleaning up one toxic dumpsite or another, or persuading consumers to buy one type of product or another. These things can make an important contribution, but without changing the systems, institutions, and values that have caused the ecological crisis, they're unlikely to amount to very much.

The work ahead appears daunting because our systems, institutions, and values seem so permanent and unchangeable. But this is exactly what the power-holders want us to think. They want ordinary people to believe that, in the words of former British Prime Minister Margaret Thatcher, "There is no alternative" to the *status quo*. The power-holders created social systems and institutions with their own agendas in mind, so it's not in their interests to change them. Quite the contrary—it's in their interests to maintain everything exactly the way it is. But change is possible! This is perhaps the single most important lesson we can learn from previous social movements. All our social, political, and economic systems and institutions, as well as all our values and beliefs, are creations of the human mind and therefore can be changed. So this is the task before us—to change Western culture. By using a combination of strategies for social change, including the ones outlined in this book, the U.S. environmental health movement can help to do this, thereby helping to ensure the future health and well-being of our species and the earth itself.

The task of achieving environmental health for all has fallen on our shoulders, and we should not shirk the responsibility. For the sake of our children and our grandchildren, we must rise to the challenge. Environmental health may not be a goal we can achieve in our lifetimes, but it is a goal to make a lifetime. Recognizing that there is nothing more likely to succeed than a committed, widespread, and long-lasting social movement, we must try to create a healthy, just, and sustainable society. Let us be strong and, in the words of English poet Percy Bysshe Shelley,

> Rise like lions after slumber
> In unvanquishable number!
> Shake your chains to earth, like dew
> Which in sleep had fallen on you—Ye are many; they are few.[37]

NOTES

1. David Bobbit. *The rhetoric of redemption: Kenneth Burke's redemption drama and Martin Luther King Jr.'s 'I have a dream' speech.* Lanbury, MD: Rowman & Littlefield (2004).
2. Proverbs 29:18. The Bible. King James Version.
3. Lewis Carroll. *Alice's adventures in wonderland.* London, UK: The Folio Society (1961).
4. Lin Kaatz Chary. Personal communication. June 14, 2010.
5. Ottawa Charter for Health Promotion. *Canadian Journal of Public Health* 77(6): 425–430 (1986).
6. Thomas McKeown. The role of medicine: Dream, mirage or nemesis? Oxford, England: Basil Blackwell (1979).
7. Marc Lalonde. A new perspective on the health of Canadians. Ottawa, Canada: Minister of Supply and Services. (1974).
8. Robert G. Evans, Morris L. Barer, and Theodore R. Marmor (eds). Why are some people healthy and others not?: The determinants of health of populations. New York: Aldine de Gruyter (1994).
9. Robert Costanza, Bryan G. Norton, and Benjamin D. Haskell. *Ecosystem health: New goals for environmental management.* Washington, DC: Island Press (1992).
10. D. J. Rapport, R. Costanza, and A. J. McMichael. Assessing ecosystem health. *TREE* 13(10): 397–402 (1998).
11. *Ecological Medicine: A call for inquiry and action.* (February 2002). Science and Environmental Health Network. Available at: http://www.sehn.org/Volume_7-4.html#a2. Accessed March 1, 2012.
12. Ted Schettler. *Ecological medicine: Complex systems, health, and disease.* Science and Environmental Health Network. (October 2006). Available at: http://www.sehn.org/pdf/che,%20natl%20mtg,%202006,%20pdf.pdf. Accessed October 31 2012.
13. Katarzyna Kordas, Bo Lönnerdal, and Rebecca J. Stoltzfus. Interactions between nutrition and environmental exposures: Effects on health outcomes in women and children. *Journal of Nutrition* 137: 2794–2797 (2007).
14. *A common agenda for health & environment: Goals for the next generation.* Lowell, MA: Lowell Center for Sustainable Production at the University of Massachusetts at Lowell. Available at: http://www.sustainableproduction.org/downloads/CommonAgenda.pdf. Accessed March 1, 2012.
15. Donella Meadows. *Thinking in systems: A primer.* White River Junction, VT: Chelsea Green Publ. (2008).
16. Centers for Medicare and Medicaid Services, Office of the Actuary. *National health statistics group.* Available at: http://www.cms.hhs.gov/NationalHealthExpendData. Accessed October 31, 2012.

17. *The world health report 2000—health systems: improving performance*. Geneva: World Health Organization (2000).

18. *WHO statistical information system (WHOSIS)*. Geneva: World Health Organization, (September 2009).

19. Trust for America's Health. *Prevention for a healthier America: Investments in disease prevention yield significant savings, stronger communities*. (February 2009). Available at http://healthyamericans.org/reports/prevention08/Prevention08.pdf. Accessed October 31 2012.

20. Public health community announces major initiative on climate change, press release, American Public Health Association. (2007). Available at: http://www.apha.org/about/news/pressreleases/2007/climatechangeannouncement.htm. Accessed October 31, 2012.

21. U.S. Global Change Program. Draft Climate Assessment Report (2013). Available at www.globalchange.gov. Accessed February 12, 2013.

22. Bill McKibben. Global warming's terrifying new math. *Rolling Stone Magazine*. July 19, 2012. Available at: http://www.rollingstone.com/politics/news/global-warmings-terrifying-new-math-20120719. Accessed October 31, 2012.

23. See: http://math.350.org/. Accessed October 31, 2012.

24. Christopher Borick and Barry Rabe. Fall 2011 *National survey of American public opinion on climate change*. Issues in Governance Studies Number 45, February 2012. The Brookings Institution. Available at: http://www.brookings.edu/~/media/Files/rc/papers/2012/02_climate_change_rabe_borick/02_climate_change_rabe_borick.pdf. Accessed October 31, 2012.

25. Eli Kintisch. Extreme weather affecting public views on climate in US. *Science Insider*. July 18, 2012. Available at: http://news.sciencemag.org/scienceinsider/2012/07/extreme-weather-affecting-public.html. Accessed October 31, 2012.

26. Some European environmental health groups, including the Health and Environment Alliance (HEAL), are already active on climate disruption.

27. Mo McBroom. Policy Director, Washington Environmental Council. Personal communication. November 15, 2009.

28. Myles Horton and Paulo Freire. *We make the road by walking: Conversations on education and social change*. Philadelphia, PA: Temple University Press (1990).

29. Paulo Freire. *Pedagogy of the oppressed*. Harmondsworth, UK: Penguin Books (1972).

30. Sandra Steingraber. *Living Downstream: An ecologist looks at cancer and the environment* . Reading, ME: Addison-Wesley (1997).

31. Sandra Steingraber. *Having faith: An ecologist's journey into motherhood*. Cambridge, MA: Perseus (2001).

32. Sandra Steingraber. *Raising Eijah: Protecting our children in an age of environmental crisis*. Cambridge, MA: Perseus (2011).

33. Jeff Cohen. Article in the Ithacan newspaper on Feb 12 2010. Available at: http://theithacan.org/am/publish/news/201002_Ecolo-gist_speaks_about_mainstream_media_printer.shtml. Accessed March 1, 2012.

34. Parker Palmer. *Let your life speak: Listening for the voice of vocation*. San Francisco: Jossey-Bass (1999). p. 30-31.

35. World Health Organization. *The determinants of health*. Available at: http://www.who.int/hia/evidence/doh/en/. Accessed October 31, 2012.

36. John Donne. Meditation XVII. (1624).

37. Percy Bysshe Shelley. *The mask of anarchy*. (1819).

A Chronology of Key Events

1845: John Griscom published *The Sanitary Condition of the Laboring Population of New York, with Suggestions for Its Improvement*

1850: The Shattuck report on sanitary conditions in Boston was published, providing a comprehensive plan for public and environmental health management

1865: Stephen Smith and his colleagues published *The Citizen's Association Report on the Sanitary Condition of the City*

1866: New York passed the first comprehensive public health legislation in the United States, establishing the New York City Metropolitan Board of Health

1867: Massachusetts enacted the first government-authorized workplace inspections

1872: The American Public Health Association founded

1873: The Sanitarian Magazine started publication

1887:

Ellen Swallow Richards conducted the first major water quality survey in the States

Public health laboratory established in New York

1889: Hull House, part of the Settlement Movement, was cofounded in Chicago by Jane Addams and Ellen Starr Gates

1892: The Sierra Club was founded by John Muir

1893: The World's Columbian Exposition was held in Chicago, showcasing the "White City"

1901: The McMillan Commission released its plan for Washington, DC

1906: *Pure Food and Drugs Act* passed by Congress

1908:

Alice Hamilton wrote her first article on occupational health, helping to establish it as a scientific legitimate discipline in the United Sates

Jersey City, NJ, became the first city to use chlorine as a drinking water disinfectant

1910: *Federal Insecticide Act* passed by Congress

1916: New York became the first U.S. city to adopt zoning regulations

1923: The Regional Planning Association of America (RPAA) founded by architects Clarence Stein and Henry Wright

1929: Radburn, NJ—the first Garden City in the United States—was planned by Stein and Wright

1938: The *Federal Food Drug and Cosmetic Act* mandated the Food and Drug Administration to set legally enforceable food standards

1943: First recorded episode of Los Angeles smog

1945: The United States detonates atomic bombs at Hiroshima and Nagasaki

1947:

The Federal Insecticide, Fungicide and Rodenticide Act passed by Congress

California became the first state to enact air pollution legislation

1948:

The Donora, PA, air pollution episode

The *Water Pollution Control Act* enacted

1955 *Air Pollution Control Act* passed by Congress

1957 The Committee for a Sane Nuclear Policy (SANE) started to advocate for a ban on nuclear weapons testing

1958:

Barry Commoner launched the Baby Tooth Survey

The Delaney Clause of the *Food, Drug and Cosmetic Act* prohibited carcinogenic additives in food

1962 Rachel Carson published *Silent Spring*

1963 The Limited Test Ban Treaty prohibited nuclear weapons tests in the atmosphere, outer space, and underwater

1967 The Environmental Defense Fund established

1968 Paul Ehrlich published *The Population Bomb*

1969 The Cuyahoga River caught fire in Ohio

1970:

National Environmental Policy Act enacted, *Clean Air Act Amendments* enacted

The first Earth Day was on April 22

The Environmental Protection Agency created

The *Occupational Safety and Health Act* was enacted, creating the Occupational Safety and Health Administration

1971 Barry Commoner published *The Closing Circle*

1972:

The *Federal Water Pollution Control Act*, the *Federal Environmental Pesticides Control Act*, and the *Consumer Product Safety Act* were passed by Congress

First Committee on Occupational Safety and Health (COSH) formed in Chicago, IL

The Club of Rome published *The Limits to Growth*

The first Great Lakes Water Quality agreement signed by the governments of the United States and Canada

1973 The *Lead-Based Paint Poisoning Prevention Act* enacted

1974 The *Safe Drinking Water Act* enacted

1976:

The *Toxic Substances Control Act* and the *Resource Conservation and Recovery Act* enacted

Commonweal—a nonprofit health and environmental research institute founded

1977:

Further amendments to the *Clean Air Act* and the *Clean Water Act* enacted

Consumer Product Safety Commission announced a final ban on lead-containing paint

1978:

Love Canal disaster

The second *Great Lakes Water Quality Agreement* signed by the governments of the United States and Canada

1979:

Three Mile Island nuclear power plant disaster

Municipal wells in Woburn, MA, found to be contaminated with trichloroethylene (TCE) and perchloroethylene (PERC)

African American homeowners in Houston, TX, fought to prevent a garbage dump in the first environmental justice protest

Church Rock disaster in New Mexico, when a dam broke, releasing radioactive tailings and heavy metal effluent from a uranium mine

1980 The *Comprehensive Environmental Response, Compensation, and Liability Act* (Superfund) enacted

1981:

The Citizens' Clearinghouse for Hazardous Waste established

Philadelphia, PA, became the first city to enact a right-to-know law

1982:

Environmental justice activists in Warren County, NC, protest a PCB waste site

Times Beach, MO, disaster—the largest public exposure to dioxins in the United States

Superfund Amendments and Reauthorization Act enacted

1984:

The National Toxics Campaign was established
Cesar Chavez launched the Grapes of Wrath campaign

1985:

The Agency for Toxic Substances and Disease Registry (ATSDR) was formally organized
The Pesticides Action Network launched its Dirty Dozen campaign

1986:

Emergency Planning and Community Right-to-Know Act enacted, establishing the Toxics Release Inventory
Rachel's News started publication

1987 *The Great Lakes Water Quality Agreement* amended

1988:

The Louisiana Toxics March
West Harlem Environmental Action (WE-ACT) founded

1989:

The NRDC's report *Intolerable Risk: Pesticides in our Children's Food* released
Mothers and Others founded
Massachusetts *Toxics Use Reduction Act* enacted

1990:

The *Pollution Prevention Act* enacted
Two regional environmental justice organizations—the Gulf Coast Tenant Leadership Development Project and the Southwest Organizing Project—arranged for letters to be sent to many of the largest environmental groups accusing them of racism
Indigenous Environmental Network formed

1991:

The First National People of Color Environmental Leadership Summit held in Washington, DC
The Centers for Disease Control (CDC) reduced its action level for lead in blood from 25 micrograms per deciliter to 10 micrograms per deciliter

1992:

Children's Environmental Health Network established
Deep South Center for Environmental Justice founded

1993:
Environmental Working Group founded
Science and Environmental Health Network founded

1994 Executive Order on *Federal Actions to Address Environmental Justice in Minority Populations and Low Income Populations*

1995 Women's Voices for the Earth founded

1996:
Our Stolen Future published
Health Care Without Harm established
Food Quality Protection Act enacted, repealing the Delaney Clause
The *Clean Air Act* banned the sale of leaded gasoline for on-road uses

1997 President Clinton signed an Executive Order on children's environmental health

1998:
Wingspread Conference on the precautionary principle
Environmental Health Fund founded

1999 Institute for Children's Environmental Health and the partnership for Children's Health and the Environment founded

2000 The *Children's Act* authorized a major study on the effects of the environment on children

2001 The Stockholm Convention on Persistent Organic Pollutants
Coming Clean formed

2002:
The Collaborative on Health and Environment established
Environmental Health News started to publish *Above the Fold*, a daily e-news service

2003 San Francisco makes the precautionary principle the basis for its environmental policy

2004:

The *Louisville Charter for Safer Chemicals* agreed
Investor Environmental Health Network formed

2005:
SAFER States founded
California's *Safe Cosmetics Act* enacted

2006:
The Environmental Justice For All Tour
Greenpeace launched its *Green My Apple* campaign
BlueGreen Alliance formed
Business-NGO Working Group for Safer Chemicals and Sustainable Materials established

2007 REACH entered into force in Europe

2009:
Safer Chemicals, Healthy Families coalition launched
The ATSDR launched a *National Conversation on Public Health and Chemical Exposure*

2010:
The *Safe Chemicals Act* introduced into Congress to replace the *Toxic Substances Control Act* (1976)
The *Safe Cosmetics Act* introduced into Congress
San Francisco passes an ordinance requiring retailers to provide information on how much radiation is emitted by the cell phones they sell

2012:
The Nuclear Regulatory Commission approved a new nuclear power plant in Georgia— the first since the Three Mile Island disaster in 1979
EPA released the first volume of its dioxin reassessment, more than twenty-five years after its original assessment
The Centers for Disease Control recommends a "reference value" to identify children with elevated blood lead levels
Amended *Great Lakes Water Quality Agreement* was signed by the governments of the United States and Canada

Selected Environmental Health Resources[1]

NATIONAL GROUPS

Center for Environmental Health: www.ceh.org
Center for Health, Environment & Justice: www.chej.org
Children's Environmental Health Institute: www.cehi.org
Children's Environmental Health Network: www.cehn.org
Clean Production Action: www.cleanhealthyny.org/
Collaborative on Health and the Environment: www.healthandenvironment.org
EcoMom Alliance: http://www.ecomomalliance.org/
Environmental Defense Fund: http://www.edf.org/
Environmental Working Group: www.ewg.org
Health Care Without Harm: www.noharm.org
Healthy Child Healthy World: www.healthychild.org
Healthy Schools Network, Inc.: www.healthyschools.org
Institute for Agriculture and Trade Policy: www.iatp.org
Natural Resources Defense Council: www.nrdc.org
Pesticide Action Network of North America: www.panna.org
Physicians for Social Responsibility Environmental Health Policy Institute: www.psr.org/environment-and-health/environmental-health-policy-institute
Science & Environmental Health Network: www.sehn.org
Silent Spring Institute: www.silentspring.org
Women's Voices for the Earth: www.womensvoices.org

STATE/REGIONAL GROUPS

Alaska: Alaska Community Action on Toxics: www.akaction.org
Connecticut: Coalition for a Safe and Healthy Connecticut: http://safe-healthyct.org/
California:

Environmental Health Coalition:www.environmentalhealth.org
Environmental Health Network: www.ehnca.org
Californians for a Healthy & Green Economy: http://www.changecalifornia.org/

Great Lakes: Great Lakes United: http://www.glu.org/
Maine: Environmental Health Strategy Center: http://www.preventharm.org/
Massachusetts: Alliance for a Healthy Tomorrow: www.healthytomorrow.org
Michigan: Ecology Center: www.ecocenter.org
New York:

Cancer Action New York: www.canceractionny.org
Citizens Environmental Coalition: http://www.cectoxic.org/
Clean and Healthy New York: http://www.cleanhealthyny.org/

Oregon: Oregon Environmental Council: www.oeconline.org
North Carolina: Toxic Free North Carolina: http://www.toxicfreenc.org/
Washington: Washington Toxics Coalition: www.watoxics.org

LOCAL AND ENVIRONMENTAL JUSTICE GROUPS

Communities for a Better Environment: http://www.cbecal.org/
Community In-Power & Development Association: http://mycida.org/
Connecticut Coalition for Environmental Justice: http://www.environmental-justice.org/
Deep South Center for Environmental Justice: http://www.dscej.org/
Honor the Earth: http://www.honorearth.org/
Indigenous Environmental Network: http://www.ienearth.org/
Inter-Tribal Environmental Council: http://www.ntec.org/
Just Green Partnership: http://just-green.org/
Just Transition Alliance: http://www.jtalliance.org/
National Tribal Environmental Council: http://www.ntec.org/
Southwest Network for Environmental and Economic Justice: http://www.sneej.org/
Sustainable South Bronx: http://www.ssbx.org/
West Harlem Environmental Action (WE ACT): www.weact.org

ORGANIZATIONS REPRESENTING PEOPLE WITH ENVIRONMENTALLY RELATED DISEASES

American Association on Intellectual and Developmental Disabilities: http://www.aaidd.org/
American Lung Association: http://www.lung.org/
Autism Society: www.autism-society.org
Birth Defect Research for Children, Inc.: www.birthdefects.org
Breast Cancer Fund: www.breastcancerfund.org
ElectromagneticHealth.org: electromagnetichealth.org
Learning Disabilities Association of America: http://www.ldanatl.org/
MCS Survivors: http://mcsurvivors.com/
Parkinson's Disease Foundation: www.pdf.org

BOOKS

A Community Guide to Environmental Health, Conant, J., Fadem, P., Hesperian Books: 2008.
A Small Dose of Toxicology: The Health Effects of Common Chemicals, Gilbert S. Institute for Neurotoxicology and Neurological Health: 2004.
Body Toxic: How the Hazardous Chemistry of Everyday Things Threatens Our Health and Well-Being, Baker, N., North Point Press: 2008.
Chasing Molecules: Poisonous Products, Human Health and the Promise of Green Chemistry, Grossman, E., Shearwater: 2009.
Diagnosis: Mercury, Hightower, J., Island Press: 2009.
Having Faith: An Ecologist ' s Journey into Motherhood, Steingraber, S., Perseus: 2001.
Living Downstream: A Scientist ' s Personal Investigation of Cancer and the Environment, Steingraber ,S., Random House: 1998.
Not Just a Pretty Face: The Ugly Side of the Beauty Industry, Malkan, S., New Society Press: 2007.
Our Stolen Future: How We Are Threatening Our Fertility, Intelligence and Survival —A Scientific Detective Story. Colborn, T., Dumanoski, D., and Perterson Myers, J., Plume: 1997.
Poisoned for Profit: How Toxins Are Making Our Children Chronically Ill, Shabecoff, P., Shabecoff, A., Chelsea Green Publishing: 2010.
Raising Elijah: Protecting our Children in an Age of Environmental Crisis, Steingraber, S., Perseus: 2011.
Sacrifice Zones: The Front Lines of Toxic Chemical Exposure in the United States, Lerner, S., MIT Press: 2010.
Sustainable Communities and the Challenge of Environmental Justice, Agyeman, J., New York University Press: 2005.
The Quest for Environmental Justice: Human Rights and the Politics of Pollution, Bullard, R. and Waters, M., Sierra Club Books: 2005.
When Smoke Ran Like Water: Tales of Environmental Deception and the Battle against Pollution, Davis. D. L., Basic Books: 2002.

NOTES

1. This list was prepared with the assistance of Nancy Hepp, Research and Communications Specialist, Collaborative on Health and the Environment.

Index

About the Author

Kate Davies is currently core faculty at Antioch University Seattle's Center for Creative Change and clinical associate professor in the School of Public Health at the University of Washington. She has been active in environmental health issues for thirty-five years in the United States and Canada. She has worked with numerous nongovernmental and governmental organizations, including Greenpeace, the Collaborative on Health and the Environment, the Institute for Children's Environmental Health, the International Joint Commission, and the Royal Society of Canada.